The Woman Said Yes

OTHERS BOOKS BY JESSAMYN WEST

The Friendly Persuasion
A Mirror for the Sky
The Witch Diggers
Cress Delahanty
Love, Death, and the Ladies' Drill Team
To See the Dream
Love Is Not What You Think
South of the Angels
A Matter of Time
Leafy Rivers
Except for Me and Thee
Crimson Ramblers of the World, Farewell
Hide and Seek
The Secret Look
The Massacre at Fall Creek
Double Discovery
The State of Stony Lonesome

JESSAMYN WEST

The Woman Said Yes

Encounters with Life and Death

MEMOIRS

A Harvest/HBJ Book
Harcourt Brace Jovanovich, Publishers
San Diego New York London

Requests for permission to make copies of any
part of the work should be mailed to:
Permissions, Harcourt Brace Jovanovich,
Publishers, Orlando, Florida 32887.

Library of Congress Cataloging in Publication Data
West, Jessamyn.
The woman said yes.
CONTENTS: Grac.—Carmen.
1. West, Jessamyn—Biography. I. Title.
PS3545.E8315Z518 813'.5'4[B] 75-40368
ISBN 0-15-198400-X
ISBN 0-15-698290-0 (Harvest/HBJ : pbk.)

Printed in the United States of America

First Harvest/HBJ edition 1986

A B C D E F G H I J

With love for Grace and Carmen
and to celebrate their courage

Actually, there were two women,
possibly three. One woman, however,
was the mother of the other two,
and her "yes" lay behind all
their subsequent affirmations.

Part One

GRACE

THE WOMAN who said yes had four children, two of them girls. The woman's name was Grace Anna and she hated it. She wanted to be called Gladys Juanita, a name she considered romantic. A Quaker, born in the backwoods of Indiana, her own name, amidst the Hannahs and Rebeccas and Janes, *was* romantic; but she longed for something lusher. There were plenty of odd names there: Missouri Overturff, Opalescent Treadway, Spicey Platter. But oddness wasn't what she had in mind. Mother (Grace Anna was my mother) wanted a name that suggested scenes far removed from the cornfields and wood lots, from the pawpaw and persimmon patches of Jennings County. She wanted a name that reflected the dream in her mind.

This dream came no doubt from her reading: Shakespeare, *The Life of Queen Victoria* by the Reverend John Rusk, Mary J. Holme's *English Orphans,* and *St. Elmo* by Augusta Jane Evans Wilson. In this dream, girls did not churn butter or bring in the eggs. They did not wear sunbonnets or shimmies. Instead, modishly dressed, they sat in the drawing rooms of English manor houses and carried on conversations with Wilfreds and Gareths and Marmadukes. So Grace, with this dream in her mind, gave her two girls (her boys also, but that is another story) names suitable for such settings and such conversations.

I, her first born, was named Jessamyn. Where she got the first half of this name, I know. I was named, at my grandmother's insistence, for my grandfather, Jesse. Now, in my mother's dream, no girl named Jessie sat in a wing-back chair and talked about *Ivanhoe* with a Marmaduke. Jessies wore sunbonnets and gathered the eggs.

Grace, until she met Eldo, who was to be her husband, thought that her father, Jesse, was the best man on earth. But a daughter who should surely be nicknamed Jess didn't fit the dream in her head, so I was called Jessamyn. This made a bow in Father Jess's direction without tying me down to egg-gathering and sunbonnet-wearing for the rest of my life.

Her second daughter (Grace had taken care of namesake duties to her own parents and to Eldo's by the time this daughter was born) was called Carmen. Where Grace got this name, I know not. It didn't suit the English-manor-house dream, unless, like Vita Sackville-West's Spanish mother, Carmen was to marry an English milord and settle down in a house like the Sackvilles' Knole. But whatever the reason, Carmen she was named.

I do not mention my brothers' equally exotic names. As I said, they are not a part of this story; and, besides, neither was ever called by his given name. The older, because of his white curly hair, was always called Cotton. The younger, because of his red curly hair, was, and still is, called Rusty.

Carmen and I did not have nicknames. We were towheads, gradually darkening. I, in particular, was told I must never, never permit my name to be shortened to Jess; and I never have. No one ever wanted to nickname Carmen, so she had no problem.

The three of us grew up together: Grace, Jessamyn, and Carmen. But this is chiefly Grace's story—the story of a woman who said yes to the life in her own life, and whose efforts and example helped her daughters to say yes to the life in their lives, and courageous enough when one life no longer held life to say yes to death. The second part of the story is Carmen's. I was with both, and I am still alive and hence tell their story.

Grace is dead now. When a person dies, lovers lament the loss of flesh—and soul, too, I suppose. But what the world loses when a human being dies are the dreams he has had and still possibly has in his heart. They, these dreams and imaginings and hopes, constitute a great, many-tiered structure, more spacious than Babylon, more beautiful than Acoma. These dreams, as compared with the bones and tissues of the body, are as multitudinous as the stars in the galaxies that populate the heavens over our planet.

When my mother died, I missed her physical presence. But what I thought then and still think about is the subtraction from the world of her dreams. I do not say "loss" because the world knew nothing of these dreams; knew no more of them than the pretelescopic world knew of those galaxies that endlessly, insofar as we know, encircle us.

I was told many of these dreams. When I look at Grace's earliest pictures, I can distinguish them behind the four-year-old brow and lurking in the corners of her solemn mouth.

Grace Anna McManaman Milhous was born in a log cabin. When she was eight, a photographer came out to take a picture of the family as it posed on that vine-covered front porch. Beside Grace stands Walter, her lanky ten-year-old brother; next are her father and mother; also in the picture is the hired man, a fine-looking dandy with a great roach of shiny black hair; and the twelve-year-old boy, brought from the orphan asylum in Indianapolis to work for his bed and board, is there. The family pets are there, too: a starling in a wooden cage, tongue cut so that it could speak; a large dog, breed unknown, with long black curly hair. His *name* I know. He was called Old Pedro.

It must have been summertime. The children are barefooted. The vine over the porch—clematis? honeysuckle?—is in bloom. I have had the picture enlarged, and what I took to be large blossoms of some sort in the grass, peonies, perhaps (always called pineys there), turn out to be hens, strangely white; strange because I thought that Plymouth Rocks, Buff Orpingtons, and Rhode Island Reds were the breeds common at that time.

Thus it was on a midsummer morning (the light in the picture falls like summer morning light) in the 1890s in southern Indiana.

Photography was a marvel then. Families, instead

of the moon, were, in the 1890s, the great pictorial subject. Children were thought as photogenic as craters. There is another picture of Grace Anna, aged four, with her brother, Walter, aged six. Grace holds in her arms a doll. Her soft, childish face is that of a mother bereaved; eyelids swollen, mouth drooping. But for the sake of the child who remains, she is defiant. She had come to the photographer with two dolls: Lily White and Rose Red; she wanted both in the picture, but Brother Walter had said that he'd be shot before he'd have his picture taken with a girl holding two dolls. So one doll, Lily White, I believe from the black ringlets of the doll Grace is carrying, was discarded. A good choice, in my opinion; but on the child's face is the sorrow of a mother who has had to make this hard decision. Four years old, but the dream already is of motherhood.

In the next picture, Grace is fourteen. She has borrowed her mother's wedding ring, but the photographer had told her that it would not be seemly for an unmarried girl of fourteen to wear a wedding ring. He had placed her unbanded hand on top of the hand with the ring, to conceal the ring. But Grace was too shrewd for a small-town photographer. The minute his head went under the camera's dark cloth, she changed hands. In the picture she sits, wedding ring visible on wedding finger; the dream of wifehood showing in her face.

In the final picture I have of Grace as a girl, she is fifteen. She stands in the back row of the graduating class, eleven girls and three boys, of the North Vernon Normal School. All of them have been, after a six weeks' course, certificated by state and county to teach any or all of the classes of a grade school. Grace is the only graduate without a face. She has scratched her face away. One day, having a good square look in a mirror, she saw that the dream she had always had of being a pretty girl was a dream only. She was not pretty. She was a plain girl. Her face was long, her nose large, her eyelashes short; her skin in summer so freckled that she was called "turkey egg."

From that time on, Grace, in her pictures, while she never again scratched her face away, is always shadowed by a large hat; or she is standing behind a taller person; or is laughing so that she is mostly white teeth and half-closed eyes.

What she had that the cameras couldn't catch, nor she either when she peered into mirrors, was something that flared up from inside her: lit up the big blue-green eyes, curled the big soft red mouth, and enlivened the thick reddish hair.

Not red. It truly was not, and even if it had been, in those days red was not a color to claim. Red hair was a sign of Irishness, and to be Irish in those days was not, people thought, a matter to boast about. The Irish drank, were lazy, and lived in shanties. They were also Papists. A Papist could sin today, confess tomorrow, and be ready with a clear conscience to tip the bottle, kiss the girls, and play a card off the bottom of the deck the minute his penitential prayers were heard. The Irish themselves did not care for red hair. There was no record of a saint, they said, with red hair.

Grace *was* Irish; all the more reason, perhaps, she resented until her last years any reference to reddish lights in her hair.

She had other dreams, of course, than those that these pictures suggest.

It was the practice of Hoosier girls to duel in the spring with Johnny-jump-ups. Each girl would select a stout specimen (Johnny-jump-ups are wild violets), and with stem in hand attempt to hook off the head of an opposing Johnny held in the same manner by another girl. One flower always, and sometimes two, lost their heads in these contests. But afterwards, regarding fallen heads, Grace dreamed that had circumstances been different, she might have dealt with humans in exactly the same way. Wilting violet heads, bleeding human heads. Cruelty started somewhere, didn't it?

Grace dreamed that her mother, quick of tongue and hand, her house cleaned and her husband and children instructed, might take it into her head to visit relatives in Kansas, leaving Grace to keep house for her father and brother. This dream, realized, would have resulted in a nightmare. Grace's mother never permitted Grace to make a bed, sweep a floor, or wash a dish. She got in the way, Grandma Mary Frances said. Grace talked with the hired girls (hired men, too) and slowed them down. She let a hen who seemed to want to lay an egg come into the house and do so. Next thing there would be pigs in the parlor and guinea hens on the whatnot. Grandma Mary Frances never went to Kansas. As a result, Grace, her father, and brother had well-cooked meals served in a clean dining room: celery in the celery-holder; spoons in the spoon vase; toothpicks in the toothpick jar; and a small saucer placed beside the big saucer to catch the drip, if there was any, from the coffee cup.

Grace, though she was a great reader and a voluminous letter-writer, did not dare dream of anything so presumptuous as formal writing. She had never seen a writer; no one she knew had ever seen a writer. The only writers she had ever seen were those encased in oval frames and covered with glass. All, with the exception of Emerson, had long beards. Their pictures hung on the walls of the one-room school she attended at Fairmont.

What she could do, and what everyone except those who stuttered or boys who thought it sissy did do, was recite. In this they were like musicians: unable to compose music, they vocalized. Grandma, who believed in doing things right, saw to it that Grace, who liked to recite, had elocution lessons. Parents, then as now, took pride in their children's public performances. There was then no tap dancing, no ballet, but when a daughter recited "The Assyrian came down like the wolf on the fold," she was a surrogate for Mother, who, if *she* had had elocution lessons, could have done as well.

Grace did very well indeed. She won prizes: globe-shaped paperweights of glass with multicolored fountains of glass spraying upward inside; cream pitchers shaped like cows, with the tail in a curve for a handle and the mouth wide open for a spout; autograph albums of red Leatherette embossed with gilt print.

Did Grace dream of being an actress? I think not. How could she? She had never seen a play. No pictures of Anna Held or Maude Adams hung on the walls of the Fairmont School. I do not think she even dreamed of being an elocution teacher. She *was,* however, when she recited her poems, already an actress.

She still recited the poems of her youth when I was a child. (What do I mean, "youth"? Grace was twenty-five when I was six.) Listening to her, I was *in* a theater. The plays were not the world's greatest plays, but I did see the performances of a skilled actress.

I can still remember "Only three grains of corn, John, only three grains of corn." A mother, her three children starving upstairs, speaks to their father. A wealthy woman has asked for one of the children. That child will be well-fed, loved, cared-for. "I looked at John, John looked at me. Which of the three shall it be?"

Grace, as she recited this poem, would glance at me (playing the part of the silent John), then cast her tear-glazed eyes toward the bedroom upstairs. To this day, I believe those three starving children were up there. By the end of the poem the mother decided, to my great relief, that starve or not, no child would be given away.

But what I never understood then—or now—were those three grains of corn. Were they reduced to eating chicken food? I never asked, of course. The greatest of plays have in them oddities for which the greatest of performers are not responsible; and you do not spoil the performance with questions.

Grace sang—somewhat in the style of Rex Harrison in *My Fair Lady,* for she had no singing voice—"The Jealous Lover." I also remember its words: "One night when the moon was shining and the stars were shining, too, a jealous lover crept nearer to where his sweetheart lay."

When Grace sang this, I stayed away from opened windows. Before the poem was finished, the knife of the jealous lover would be dripping with the blood of his faithless sweetheart. And just as I believed in the starving children upstairs, I believed in the jealous lover prowling outside—and I did not want him, creeping nearer, to mistake me for his faithless sweetheart.

But neither starving children nor faithless sweethearts, sad as was their plight, touched me as deeply as the fate of Kentucky Belle, the equine heroine of a narrative poem bearing her name. Kentucky Belle was a beautiful thoroughbred mare, the property of a young Kentuckian. One night, Yankees, sweeping down from the north on a raid, stole Kentucky Belle. Her owner, though he searched for her for the rest of his life, never saw her again. He imagined that gentle, fleet mare frightened, perhaps wounded, during the charge at Chancellorsville, perhaps dead on Missionary Ridge. Or did she survive the war? Was she pulling a Yankee surrey to church every Sunday in York State? Or straining between the tugs of an Amish wagon? All of her beauty and hot Kentucky racing blood lost on plowmen and churchgoers? Why do children cry more, and rejoice more, for animals than for humans? *Bob, Son of Battle, Black Beauty, Rascal?*

This is not to be an account of the dreams and heart-yearnings of Grace Anna. I do not know the half of them; and there is not paper enough or time to recount even those I do know. A final dream I recount—and for it I give Grace praise.

She had taken that good look at herself in a mirror

and had seen that she was not a beautiful girl; not a pretty girl, even. Nevertheless, she did not give up her dream of finding, loving, and marrying a beautiful boy.

Most girls, after their mirrors have told them this sad story, give up their dream boys; boys both gentle and strong, black curls springing and black eyes sparkling. The age of conscious female aggression had not yet arrived when Grace was young; or at least had not yet arrived in backwoods Quaker communities. The plain girl folded her hands, minded her manners, and hoped for the best: which in her case would be a good, plain farm boy, who after they were married would give her a hand with the spring housecleaning and take a bath every Saturday night.

Not Grace. A child cannot appraise a parent—particularly in respect to what is known as "sex appeal." Grace evidently had it. Freckled, with reddish hair, not built like Lillian Russell, given to pranks (she took the slats out of the hired man's bed, and he declared that it was only by the mercy of God that he was saved from suffocation as the feather bed enfolded him). In spite of these handicaps, plain, prankish, endowed with a switchblade tongue and the downcast eyes of a conventional Quaker maiden, still the boys trooped round.

What did the boys see? Feel? Where the attractions are not conventional (beauty and availability), who can say? Some responsiveness in the female that only the antennae of boys, canted from birth in that direction, can detect?

In any case, the mirror didn't tell the whole story. It told the truth about what mirrors see. But how much *can* mirrors see? Who loves by fair or freckle?

Large or small? Straight or crooked? Mirrored people do, that's who. But the Churchills and Lincolns, the Dr. Sam Johnsons and their like, do not live by any such externals. (Of course they weren't female.) They lived, and so did Grace, by the blood in their veins and the gray matter in their skulls. And if that meant, and it did for Grace, forgetting, when she was with people, that in the mirrors of their eyes anything else was visible, she, too, was a statesman and a pundit. And a vamp.

Boys of all kinds, or all of the kinds of boys to be found in a backwoods community at the turn of the century, came to court Grace: farm hands, farm-owners, grocery-store clerks, Bible salesmen, rail splitters. They were jealous of each other. They bashed each other over the head. The grocery clerk broke Grace's parasol, which the Bible salesman was carrying. The farm hand destroyed the pictures the farm-owner had had taken at the county fair of himself with Grace.

But Grace knew what she wanted, and he hadn't shown up yet. Then, one day, there he was.

On the farm next to that of Grace's parents, a nephew came to visit his aunt and uncle. He was six feet one, black-haired, olive-skinned, had broad shoulders, gray eyes, and an aquiline nose. He was, for eighteen, learned. He was reading law. He wrote poetry, sand, and made soda biscuits. (Grace didn't discover this until later.)

Feeling the beat of her pulses, responding to the click of her fast-moving gray cells, Grace decided that the neighborly thing to do was to take some cookies over to the boy's Aunt Theresa.

No point in penalizing Aunt Theresa, a kind and

neighborly woman, just because a handsome nephew was visiting her.

Grace rode bareback (horse), barefoot (Grace), with cookies (baked by her mother), through the woods to Aunt Theresa's. The horse was Old Polly, fat, gray, and twenty. Grace was fifteen, five feet four, and weighed one hundred pounds.

I don't know what Eldo's dream had been, but Grace met hers that spring afternoon. There were things about him she *hadn't* dreamed: the peculiar name; some Indian blood; his family, renters, which in that neck of the woods was at that time the counterpart of sharecroppers farther south.

Eldo was handsome enough to have his pick of girls. Would he have picked Grace? Grace didn't take any chances. He matched her dream. She picked *him.* He might never have discovered *her;* or had the nerve to ask. She was landed; he wasn't. She was already a spirited elocutionist, and his courtroom dreams were all in the future.

After Grace's death, Eldo told me that Grace had done the proposing. He had said yes at once and had never regretted it. A favorite saying of his was "As a man marries, so is he." He turned out well, and he gave all the credit to Grace. Eldo was a quiet man, easily discouraged, and given to melancholy. Grace was tenacious, witty, and merry, a Sir Thomas More of the backwoods.

An elderly woman (Grace took care of the elderly in the way others take care of stray cats) once spent the winter with her. She required care, and Grace gave it unstintingly. But when she left, Grace said, "She never made me laugh once all winter. I can't forgive her that."

• • •

I inherited some of Grace's desire to make people laugh—in my home, that is. In high school, female clowns were not highly regarded, and I knew it. There, I was a sober student. But Grace, without telling me, entered my name in a contest for Charlie Chaplin imitators. She thought that I, with cane, false mustache, and baggy pants, would be an easy winner. I might have been; and that, to my mind, would have been worst of all. Not only a clown, but a publicly branded clown. And this at a time when I wanted to be noticed all right, but not for baggy pants, mustache, and cane. I refused, of course, to compete. I hated Grace for being blind—not to my talents, I *would* have been pretty funny in baggy pants; but for being ignorant of *my* dream, which was to be Theda Bara, not Charlie Chaplin. And Grace? She had spent one dollar to pay the entrance fee to this contest, and she was exasperated with her unco-operative daughter.

But I am ahead of Grace's story. Here she is fifteen, as yet unmarried, and I have been writing about her as the mother of a teen-ager.

Grace first saw Eldo when he was eighteen and she was fifteen. Three years later they were married. Ten months after their wedding, I was born, a twelve-pound baby. The mother weighed one hundred pounds, and the birth was what they now call natural; that is, at home without anesthetics. After a two-day struggle, her hands bloody from pulling on the strips of sheeting that had been tied to the footboard of her bed, Grace produced me. I am grateful to her for her tenacity on that occasion. And she, I think, must have been glad that I came that morning unequipped with cane, pants, or even mustache.

One of the few photographs of Grace bravely facing the camera head on is her wedding picture. The wedding was in September. The bride and groom were photographed outside, under maple trees, with asters and goldenrod behind them. Grace is wearing a high-necked, gray-flecked wool dress. Not that she wanted wool or gray or a high neck. But strong-minded Mary Frances had persuaded her daughter that a wool dress, high-necked and long-sleeved, would be practical for the winter months to come. Perhaps it was. But I have been a long time sad for the white wedding dress, beautiful and impractical, that Grace dreamed of but never had.

She had something better, of course, and her face in her wedding picture showed it: she had Eldo. There is in that photograph some of the girl's spirit that cameras and mirrors did not usually catch. Marriage had not made Grace a beauty and the camera did not lie about that. It said, it says even today, that Grace on that September morning had something which if not beauty is perhaps beauty's superior—shameless love. She loves and will love for life.

She looks out over her high whaleboned collar and from under her billowing pompadour with no bridal coyness. Now that she has her own wedding ring, she makes no effort to display it. She knows that when she looks into a mirror, the face that looks back at her is not beautiful. But Eldo loves her and stands beside her. For this reason she does not try to hide behind a big grin; or to pull a spray of autumn leaves between her and the camera. What was good enough for Eldo was good enough for her.

"The children of lovers are orphans," said Robert

Louis Stevenson. Eldo begat and Grace bore four orphans.

It was not Grace's nature to let any orphan, her own or others, suffer. We knew well enough that in case of a catastrophe Grace would be there, willing to die to save her own dear orphans. Meanwhile, we were pretty much on our own. Oh, there was always food on the table (tasted pretty funny the time Grace made a mistake and put her bottle of the 1915 version of the pill in the beans); there were always covers on the beds (although if we didn't make them ourselves, they were often unmade). What was Grace doing? Out with Eldo helping furrow out; or adjusting the flow of water from the standpipes into the furrows according to the shouted directions from Eldo down at the furrow's end.

Or she was ailing. Those twelve- and fourteen-pound babies did her small frame no good. And with or without babies, the migraine that ran in the family would no doubt have plagued her. She fought it as best she could; carried on, head blistered with mustard plasters, and bound tight with whatever was at hand to keep it from bursting; the work had to be done.

When aching and vomiting, together with eyesight impaired by migraine's stained-glass-window effects, put her to bed, she went there noisily and unwillingly. She was suffering and she didn't care who knew it. I would stand at the bottom of the stairs listening for the sounds I hoped to hear coming from the bedroom stairs: groans, loud, rhythmic. Puking (a word I wasn't allowed to use then) in the bedside chamber pot. These sounds told me that Grace, though suffering, was alive. Silence frightened me.

Then I would race upstairs to ask, "Mama, are you all right?"

"No, I'm sick as a dog. But it's easing up some, I think."

"Would a hot towel help?"

"It wouldn't do any harm."

I tore downstairs to put the kettle on.

So between her love for Eldo and her own poor health, the four orphans she had borne were never dampened by Grace's tears; never heart-stricken with remorse for some real or imagined slight given their mother; never burdened by her enfolding arms. We went our own ways.

Sicknesses of our own, which might have disabled us had they been fussed over, we threw off lightly. Measles, the only childhood disease I had, went unnoticed until I jumped out of bed and fainted. It was an experience I enjoyed, but one I have never been able to repeat.

As orphans, we read what we pleased. Nothing very stirring was available. There was sex in Elinor Glyn's *Three Weeks,* I now know. But I didn't know it then. Sex was something we didn't understand; but cruelty, with which Charles Dickens' *Child's History of England* is filled, we understood instinctively. Eyes blinded, red-hot pokers rammed into human apertures, a child understands all this without explanation. If red rose petals for a bed was sex, we preferred mattresses.

We were odd dressers. I wore a dress to school wrong side out. I didn't notice it; Grace didn't notice it. Teacher did, of course.

It was Eldo, not Grace, who went with me to help pick out my high-school graduation dress. We chose a beauty, white net with pleated flounces. Neither of us at the time of the purchase had foreseen that on the day before graduation I would sunburn myself lobster-red with the senior class at the beach. At the graduation exercises, I appeared to be wearing long flannel underwear. Or to be the only redskin graduating. Grace thought the effect interesting. "You put everyone else in the shade," she declared. I at least put everyone else in mind of the value of shade at the beach.

After First Day Meeting was over, my older brother and I always set off for an afternoon hike through the cactus-covered foothills, which rang with what sounded like rattlesnakes and sometimes were; usually the dry rattle came from the cicadas' afternoon concerts. We killed rattlesnakes when we saw them. Cicadas we knew were harmless, and we waded through them, skin twitching for fear they *might* be snakes, fooling us. We peeled and ate cactus apples; and when they were ripe, we ate elderberries.

This didn't prevent us from coming home after our ten-mile hikes starved. In gravy bowls, we mixed what we called "cocoa mud," a mixture mud-thick and mud-colored of cocoa, sugar, and evaporated milk. Some mothers would have been frantic about those days of cactus, rattlesnakes, and cocoa-mud. Not Grace. She thought her orphans were smart: they could avoid snakes and digest anything.

In turn, Grace's children knew that in any emergency she would be with them and for them. They believed that this woman who could not swim a stroke would conquer a rip tide to rescue them, were they carried out to sea. They believed that this woman who feared mice and did not even like a cat to look her in the eye would strangle a mountain lion barehanded if necessary to save them. Of Eldo they expected the possible; from Grace they expected the impossible.

Grace was never required to do the impossible for them. They lived in a land where there were more ground squirrels than rattlesnakes, more road runners than people. The children, from their earliest years, had been permitted, even required, to rely on themselves. Grace expected from them the same kind of endurance she had. Groan and forget it.

Three hundred years of Quakerism had perhaps bred aggression out of them. Still, they were, on their mother's side, Welsh and Irish; on their father's side was that strain of Comanche; and none of the three noted for Quakerliness. Nevertheless, the boys stayed out of jail and the girls "out of trouble," as euphemism had it in those days. By the time she was forty-eight, Grace's three oldest were married; only one, a twenty-year-old son, was left at home.

That summer, the first of her children's calamities struck her. I, her eldest, home from college, where I had been working for my doctorate in English, had a lung hemorrhage.

I had been ailing all year and had made many trips to the university's infirmary. I was afraid that I was a hypochondriac. Grace didn't think it a sign of health to be coughing up teacups of blood. We went to the family doctor that afternoon, Grace driving an old square-nosed Reo.

The good doctor (he was a good one), named Thompson, knew at once what ailed me, though he didn't tell me until after he had seen the X-ray.

He then took me and Grace into the hospital board room. Why there, I do not know. It was a room for a hanging judge, dark, formal, and gloomy. Suitable, I admit, for the verdict he was about to deliver. Cool, too, which on a burning August afternoon was a comfort; though with my constant fever, I was more aware of air that was chilly than of air that was hot.

I was as ignorant of tuberculosis as of elephantiasis. I believed it had died out with the plague. It was difficult for me to think of it as a sickness. It was more as if I had been told I was a leprechaun. Or had second sight. And thinking back to that period now, perhaps, for a while, I had something near to that.

In the year before the verdict of T.B., I had, in spite of a doctoral candidate's heavy load of reading, Old English, Middle English, and modern French, read every account I could lay my hands on about consumptives: novels, journals, biographies, letters, autobiographies. I didn't for a minute think I had tuberculosis. I had been visiting the university's doctor for female students—a female herself, and, until recently, a librarian. I had afternoon fever, pleurisy, a dry cough—but also a negative skin test. A body (neither she nor I knew it) already combating massive inroads of tubercle bacilli ignored a pinprick of bacilli in the arm. The body operates like a competent general; it fights the main invasion and ignores peripheral skirmishes.

I had had almost no experience with doctors. A primitive's faith in his witch doctor was negligible compared with mine in a university M.D. She said that I obviously had something wrong with me: a tropical fever?, a tooth infection?, a student's anxiety? But because of the negative patch test (no X-ray was made), T.B. was ruled out.

Nevertheless, some compulsive conviction deep in my unconscious kept me reading books of or about the tubercular. Thoreau's *Journal.* (Thoreau had little to say about dying; as his life ebbed, he continued to make new discoveries—the life there was in kittens, for instance.) D. H. Lawrence's *Letters.* Jeffries' *The Story of My Heart.* And a dozen others of less lasting fame.

What drove me to read thus widely and intensely of phthisis? I do not know.

What caused me to write in my journal at that time, "For thousands of years people have been confronted by the necessity of dying. Innumerable persons have had to watch a dated death approach slowly. They have had to recognize that the very body which contained, fold on fold, the elements of their personality has turned against itself. Yet in spite of the problem being not only universal but immemorial, little has been written on the subject. We have bookshelves filled with tomes telling us how to achieve or endure puberty, marriage, motherhood, and what is known as 'the change of life.' But we have no handy guides to dying. There is a book called *Holy Living and Dying,* but it is more praised for its literary qualities than its practical hints.

"Of the many who have walked to that black gate, a few have touched it, then for one reason or another returned. Let them advise us. What will give us the greatest comfort on our trip down that bleak road? Are there vantage points from which we can have backward glances to the sunlit land we have left behind? Or best keep steadily on? Are there shortcuts? And what of them? If the journey proves too painful, should we take them? Are we fools to plod on? Or are such shortcuts, like many others, not timesavers in the long run?

"There is great need for such a how-to book. The publishers should seek out authors capable of writing such books. Such books would have enormous success, for while marriage and motherhood are a part of the lives of the fortunate, death can be confidently expected by even the most pessimistic."

This was written two years before Dr. Thompson, in the board room, told me what *I* could expect.

Two months earlier I had copied this from Jeremy Taylor into my journal:

"Death meets us everywhere and is procured by every instrument and in all chances, and enters in at many doors, by violence and secret influence, by the aspect of a star, and the stink of a mist, by the emissions of a cloud and the meeting of a vapour, by the fall of a chariot and the stumbling at a stone, by watching at wine, or watching at prayers, by the sun or the moon, by a heat or a cold, by sleepless nights or sleeping days, by water frozen into the hardness and sharpness of a dagger, or water thawed into the floods of a river, by a hair, or a raisin, by violent motion or sitting still, by severity or dissolution, by God's mercy or God's anger."

So for two years death had been on my mind, and death of a particular kind; though neither I nor the university doctor had a name for my disease. It took a teacup of blood to identify my trouble. Keats, more than a hundred years earlier, had been smarter. "That is arterial blood," said he at the sight of the first stain. Doctors then, who knew tuberculosis when they saw it, treated his lung hemorrhage as an indication of an oversupply of blood. He was further bled and put on diets of dry toast. Keats with the stamina of a mule, and who would have survived his disease, could not survive his doctors. Treated by them, he was dead at twenty-five.

Dr. Thompson held my X-ray up so that Grace and I could see it.

"We find here," he said, "irregular infiltration throughout the upper and medial portion of both lungs, with cavitation in the apical portions of both lungs. The findings are those of a far advanced bilateral tuberculosis with considerable fibroid reaction."

"You mean she has consumption," said Grace.

"Yes, Mrs. West. A well-developed case of pulmonary tuberculosis. Pulmonary, meaning 'of the lungs.'"

"I guessed that," said Grace.

Grace said to me, but in a voice that could be overheard, "He couldn't speak with more satisfaction if he had developed it himself."

Dr. Thompson gave Grace a hard look. She was his own patient. He had delivered her last baby; but he had never been able to do anything about teaching her a sickbed manner the equal of his own bedside manner.

"Now, Mrs. West, let us outline a program for your daughter."

Grace said nothing.

"She will have to go to bed for at least two years. Preferably in a sanatorium."

Then he turned to me. "But you must not think

your life is over. Some of the most important work in the world has been done by invalids and invalids with your disease. Keats, for instance, wrote some very fine poems after the onset of his phthisis. Very fine indeed. He didn't let sickness slow him down."

Dead at twenty-five, I thought.

"And Marie Bashkirtseff had a picture hung in the Luxembourg after she became sick. Kept right on painting and didn't know the meaning of the word 'defeat.'"

Dead at twenty-four.

"In music, there was Chopin. I'm no judge of music, but those who are consider him first-rate. Tuberculosis didn't stop him from composing."

Dead in his thirties.

"Then there was Farrell Loomis, a young lawyer right here in this town, who developed your trouble. He was legal adviser to the First National. They tell me he saved them fifty thousand dollars even after he'd lost one lung."

Grace got to her feet. "He died last week."

"I know. A pity. Fine young chap to pass on so early."

Grace was mad, and let the doctor know it. I felt ironic and kept my mouth shut. For whom could I save fifty thousand dollars? What anecdotal peg would I provide the doctor after I passed on? Would it be vulgar to say, "Doctor, must I die?"

Grace wasn't thinking about dying. She was thinking about recovery. "Which sanatorium, Doctor?"

"I'll give you a list."

He wrote them out. "The sooner she is in one of these and in bed, the better."

"She'll be there in three days."

I was. One cup of blood separated me from being a Ph.D. candidate and made me a certified T.B.-er.

Grace put me to bed at once like a patient with a heavy cold; covered me with blankets and fed me chicken-fried steak and milk gravy.

After the lights had been turned out at nine o'clock (anything a sanatorium could do Grace could do better), I went outside to walk barefoot onto the lawn under the August stars.

I have never been more elated. After two years of being unable to climb stairs without panting; of pleurisy pains that made me crawl to the bathroom; of cheeks flushed to plum color in the afternoon, I now had a name for these symptoms. They were not imaginary. I was not a malingerer. I was not a hypochondriac. I was like a person who, having been living in a dream of insanity, wakes up to find himself sane. Sick, yes; but sane.

For two years I had walked about, told by the doctor that nothing ailed me; as miserable with the conviction that I suffered delusions as I was with my physical misery. Now I had been declared sick, but sane. Those August stars dripped honey-easement; the grass under my bare feet was cool with dew. Instead of feeling, as would seem natural, depressed by the verdict I had just heard, I felt instead the joy of one who has been reprieved. All that was wrong with me was that I had a lung ailment.

Keats had had poor doctoring; Marie Bashkirtseff had kept on painting; Loomis had worked to save the First National fifty thousand dollars.

It would be different with me. I would have expert medical care; not paint; not work for the First National. I would take to my bed, breathe carefully, eat every mouthful put before me, and in three months, I would be well. I was different. The misfortunes of others would teach me a lesson. No one in the state of California went to bed happier than I that night. Now that I knew what was wrong with me, I would soon be well.

In three days, as she had promised the doctor, I was in a sanatorium: a sprawl of wooden buildings on a parched foothill of the Sierra Madre, just outside of Los Angeles. Below us were the highways filled with cars carrying people: people going to work in the city; going to the beaches beside the ocean glitter that we could see from our beds; going to the summer sales at Bullock's and Robinson's. Did such a world still exist? Had we once been a part of it? Had we come and gone as the fancy took us? No temp stick at five in the morning, breakfast at six, bed bath at ten, rest period (what had we been doing all day?) at two, supper at 5:30, lights out at nine?

I was put in a terminal ward. That is where hemorrhaging patients *are* put. Others are there, too. There are a variety of ways in which T.B. can kill you. The bacilli can attack stomach or throat or bladder. Or the bacilli can simply destroy the lungs, flood the system with poisons and send the temperature up and up until what lies on the bed is nothing but bones, covered with white skin (the cheeks are not red when the fire is white-hot), and eyes big as half dollars and blazing with their terrible knowledge. The bleeders often look very well. By chance their lesions are deep and near veins. They can eat a good dinner, watch *Gunsmoke,* and be dead by morning, their beds stripped and mattresses out to air. It happened in the short time I was there. For the time was short.

Terminal! Grace had no intention that the life of any child of hers should terminate at the age of twenty-eight. This being true, why put the child where what she had to see and hear would hasten what she had come to the san to prevent?

I didn't tell Grace about my first evening in the terminal ward, whether because I couldn't believe it myself, or to save her pain. The sick soon come to understand that they live in a different world from that of the well and that the two cannot communicate.

What happened was not really shattering, except in what it told me about the world I was going to live in for the three months it took me to get well.

First of all, I was given a little booklet called *What You Need to Know About Tuberculosis.* Most of what I needed to know I was learning too late; and some of what I learned I would have been better off never to have known.

"Ninety-five percent of those who enter sanatoria with far advanced tuberculosis are dead within five years." I was "far advanced" and that knowledge did nothing to speed up my cure.

After I had read the booklet, the head of the sanatorium, a Dr. Schultz, came in. Dr. Schultz was a former army doctor, iron-gray, made of iron, maybe, with an anvil-shaped face and wandering hands. He left the army, perhaps for lack of the landscape he preferred to explore.

After greeting me, he took my hand in one of his

while with the other he spanned my arm from wrist to shoulder. "Not very emaciated yet, are we, Mrs. McPherson?"

The only answer I could think of to that was the obvious, "Give me time."

The doctor laughed. His teeth were real as tombstones and almost as large.

Next, in order for me to have a better look at my X-ray, which I didn't want to see and didn't understand even when it was explained to me, Dr. Schultz leaned over to pull down the blind that covered the window at the head of my bed. To do so, he balanced himself with one hand on a breast. For two years, in different rooms, he did this balancing act. Instead of saying, "Take your hand off me, you cad," I acted as though the positioning of his hand were happenchance (perhaps it was); and he acted in the same way. Such a touch from some hands might not have been objectionable, but there is no real pleasure to be had in being used as a handrail to expedite movement, even by physicians more appealing than Dr. Schultz.

Why didn't I object? Was I saying, "You do not exist for me"? Or was I thinking, "You have enough troubles as it is. A little handrailing won't kill me"? Or had I, as Norman Mailer says of Arthur Miller, "the psychology of poverty where the mark of a man is to suffer and endure"?

I do not know; but I do know that on my first evening in the sanatorium, "95 percent dead in five years," "not much emaciated *yet,*" breast as indifferently used as a ball bearing in a bicycle transmission, the euphoria of the night on the grass under the stars dwindled sharply. I still preferred sickness to insanity, but sickness (though I still expected to be out in three months) was not going to be a lark either. The faces around me told me that. I had seen Keats's death mask; but I had never seen those taut lips part to use a sputum cup, or heard the rales of pitted lungs.

I told Grace nothing of any of this. Some of it she saw at once for herself. I spent only one night in the terminal ward. Dr. Schultz was, or had been, an army colonel. Grace, in an emergency, was a battle general. What went on between the two I was never told, but I judge that Schultz recognized a Patton when he saw one.

His only reference, if that was what it was, to the meeting was a sour question to me: "Are you an only child?" Perhaps he thought that so much energy and concern would not be available to a woman with more than one child to battle for.

"No."

"The youngest?"

"The oldest of four."

This information fit none of his army-medical psychology. He ignored it. But I was put in a room adjacent to, though not a part of, the terminal ward, because his army-medical psychology recognized Pattons.

This irritated Dr. Schultz. Life was no doubt made easier for him by those patients who talked, with the optimism characteristic of the tubercular, of the homes to which they would soon return and of the careers which had been only temporarily interrupted. I never permitted myself such talk. I have always thought it bad luck to count unhatched chickens. True, I expected to recover in short order; not because I was consumptive but because as my mother's daughter I believed (at that stage of my illness) that it would be weak to submit to anything as minuscule as a tubercle bacillus. And I was young enough to believe that death was something that happened only to other and older people.

True, twenty-four hours in the terminal ward had been enough to show me that death doesn't spare *all* of the young. It would possibly take more efficiency than I possessed to outwit it.

Grace came to the sanatorium every other day, driving the round trip of eighty miles through rain, through Santa Ana sandstorms, through fog so thick headlights were turned on at four in the afternoon. She came with brow anointed with Musterole, with varicose veins swathed in stretchable rubber bandages, with truss laced in place. What I suffered, in actual pain, dying though I might be, was a pinprick to her multiple aches. But she came. She came laughing. She came laden with things I had never dreamed of wanting, but which proved exactly what would give pleasure to a patient next door to a terminal ward.

She came with a *papier-mâché* cow, a Jersey, I think, by its fawn coloring, which when lifted up, then put down, mooed louder than a cow whose calf had strayed. It was unbelievable. All the nurses came to see and hear the wonder. Then they took it to the terminal ward; there the moo of a cow, with the memories it evoked of pastures and wood lots that the patients had never expected to see again, made everyone forget to cough for a while.

Grace finally brought four cows. We mooed back and forth to each other, terminal to terminal, like prisoners who rap out messages to unseen cellmates. We were prisoners, and some would never be reprieved. But we remembered better times.

Grace converted a sample case, once carrying Watkins remedies or Fuller Brush bristles, which she found by the roadside, into a book carryall. It held twenty-four books, more books than the local library liked to have leave their shelves at any one time to any one person. But Grace, by mustering the cards of the whole family, had the legal means of filling her case with books. She staggered up the hill from the san's parking place, bandages, trusses, and Musterole in place, panting but triumphant with her latest finds.

She brought books I requested, books I had listed in my snobbish graduate-student way; and books I often did not know existed, but which for some reason she thought would interest me.

One book became, for as long as I was strong enough to read, the favorite reading of my T.B. days. I think I understand why now. I didn't then. The book was Scott's journal of his expedition to the South Pole. Why did this terrible account of death and disappointment so enthrall me? Well, it is, under any circumstance, an enthralling account; but consider the enticement of all that ice and snow to a person who is daily fevered. Consider the statement of a man doomed, and knowing it, "Amongst ourselves we are unendingly cheerful." And last of all consider Oates, whose frozen feet slowed down the party's progress; and who, because of this, said one morning to his comrades, "I am just going outside and may be gone some time."

Of this act, Scott wrote, "He went out into the

blizzard and we have not seen him since. We knew it was the act of a brave man and an English gentleman."

I was still of the opinion that I would not long be sick. Meanwhile, though the blizzard I had to walk through was a fire of fever, I wanted to do it as Oates had: like "a brave man and an English gentleman." I failed in this, but while I was in the san, I had Oates's example to encourage me and the Antarctic snowscapes to cool me.

One thing Grace didn't bring me was any more books by tuberculars or about tuberculosis. I doubt that I could have read them anyway. Next door to me were Keats, and Stephen Crane, and Emily Brontë themselves. You don't read about love while making love; or bone up on death while dying.

At the san (san slang was soon natural to me), "temp stick" was used for thermometer, "san" for sanatorium, "spitting rubies" or "streaking" when blood is raised with the sputum. All of the san lingo was at first repugnant to me. But to refuse to learn or use it was like refusing to learn or speak anything but English while living in a foreign land.

"You look pretty good," Mrs. McVitty, the nurse on duty, said to me on my first afternoon in terminal. "How did you happen to come up here? Been streaking?"

I know what streaking means now. I didn't know its meaning in a T.B. san then.

"Streaking?"

"Raising blood. Anything over a teaspoonful is a hemorrhage. Anything less is streaking."

"I had a hemorrhage."

"You were lucky to have it hit you early instead of late."

I didn't know for sure that it *was* early.

Mrs. McVitty, as I came to know her, was an impatient woman. She could not hide the feeling that if she were called upon to die, she would do so with infinitely more celerity and efficiency than her patients. It was a pity that the one thing they had left to do, die, they could not do well. They did not make any effort not to be troublesome.

Morning after morning, she crossed the san's threshold to find some poor skeleton still waiting to be fed and washed. She could not hide her impatience at this willful delay.

New patients brought her some respite from the monotony. To them she was an authority. They asked her questions: the significance of the morning drop in temperature? The value of pneumothorax? The danger of hemorrhaging?

She told us stories of former patients. "The last girl in this bed was just about your age. Same color of hair, too."

You must not, you must not, I told myself, but my tongue asked the question.

"Did she get well?"

"No, she passed on. She had a T.B. throat, had to whisper. She was on silence for about two months before she passed away."

My tongue spoke again.

"In this bed?"

"As a matter of fact, yes. But there's been other people in it since then."

"Are such experiences terrible to you?"

"You get used to them. My first husband died of T.B. Funny thing, in the last month of his life he had a hemorrhage every night at just twelve o'clock."

"Why was that, do you think?"

What I thought was that he was trying to get it over with. What Mrs. McVitty thought was that he always was a troublesome fellow.

"Didn't you ever take him to a sanatorium?"

"Took him to a dozen sans, but nothing helped. If you haven't got any fight in you, a san's not going to give it to you."

Didn't a sanatorium work miracles, then? I thought that I would get well, because I was I and a sanatorium was a sanatorium. But could I lie here (my hair was the same color as hers) "on silence" for two months, finally becoming completely wordless?

A T.B. san is unlike an ordinary hospital where babies are born, broken hips heal, and shotgun pellets are removed. If a man with a tail full of buckshot developed gangrene and died, it didn't mean that the woman in labor would miscarry or that the boy with his tonsils removed would bleed to death.

But in the san we all had the same disease. In a way we were all one body. If one died, each of us died a little. A bad X-ray for one was a bad X-ray for all. One patient's streaking bloodied everyone's sputum. One man's recovery gave us all hope.

A round, fat, jolly priest came to the san every week. He did visit with those of his own faith. But we all loved him, not for his religion but for his fat. He had spent six months in the terminal ward, refusing to terminate; had gradually won the battle with the bacilli and now, as rosy and bouncy as Saint

Nick, he came up the hill to the san from the parking lot, breathing easily and calling greetings as he passed our rooms, the visible sign of what could be done. He, too, had arrived "far advanced" and hemorrhaging. He knew the good he did us. He walked the halls to let us see that he was not just an "arrested" but a "cured" case. Ninety-five percent had gone underground. The other 5 percent was almost as invisible, taking care of themselves. Father Flannagan's work was public; decent public exposure.

He was our drug, and we had no others. Except for pneumothorax, the only cure then was bed rest and overeating.

Since we judged, as did the doctors and nurses, our progress by our declining temperature and increasing poundage, we all tried to be cooler and heavier.

There was not much that could be done about getting lower readings on the temp stick. The nurses were a suspicious lot, and a lower than usual afternoon reading made them suspect that a recent mouthful of water, not an abatement of fever, was the cause. The thermometer went back into the mouth, and in five more minutes the usual 101 or 102 degrees were recorded there.

Why nurses who were so alert about temperatures could be fooled so easily about weight, I don't know. Maybe they weren't fooled. Maybe they knew and said to themselves, "They have so few pleasures. Let's not deny them this."

What we did (what *I* did, and I may have been the only one) was to increase my weight by the addition of a bottle of hand lotion to my bathrobe pocket on weighing day. This necessitated additional weights each weighing day until finally I clanked onto the scales as lumpy and burdened as a pack mule.

I did not do this to fool anyone but myself; and how could I, who knew that weight lay not in any real increase of my own poundage but in jars of Pond's cold cream and bottles of Jergens lotion, find encouragement in what the scales so loaded said? I had become my own medicine man; I believed in my own witchcraft. Deprived of a tube of toothpaste or lightened by a missing bar of soap, I might have developed galloping consumption because of the weight loss.

I didn't tell Grace any of this. Of what she could really do for me, I did tell her. But she never knew about Dr. Schultz's balancing act, or the nurses' impatience, or my own insanity on weighing day. I didn't tell her that I had become afraid to make friends with the patients. The engineer from South America who waved to me. Seventeen-year-old Louise, with a red mouth and dark Valentine curls, who had enough spes phthisica to keep an entire ward hopeful; Norah Lanahan, with an Irish face, big as an acre of unplowed ground, who cried her heart out because her family had sent her money instead of gifts for her birthday. I began to know them, then they *died.* They were gone in the night; the hearse came silently up the hill at three in the morning. They left like burglars caught in the act by the law. Their mattresses were visible, airing on the balustrades of their rooms next morning.

I could endure their absence. I could not endure what their deaths did to my own confidence. Louise, Señor Padilla, Miss Lanahan: Louise was far prettier than I; Señor Padilla, far more learned; Miss Lanahan was filled with more feeling; and all as convinced as I, more so, perhaps, that *they* were not destined to be invalids or to die young. And they had *died.* Young.

I had believed that going to a sanatorium was a guaranteed cure-all. Dr. Thompson had led me to believe that. Nurse McVitty, with the story of her husband, had demolished that belief.

I had believed that if I followed every rule, ate every bite of food set before me, rested determinedly during every rest period, drank eight glasses of water a day, and breathed carefully from the stomach, not the diaphragm, I would get well.

Louise, Padilla, and Lanahan were all rule-keepers. All died. Perhaps the sanatorium saying was true: "You can keep every rule and die. You can break every rule and live. Fate decides."

Fate had perhaps already decided what was going to happen to me. Meanwhile, I might try for comfort, even if death wasn't postponed by what I did.

Sans, in the old days, were places where many, perhaps most, people died. The staff was accustomed to the discomfort and complaints incidental to this tedious process. Grace wasn't. She didn't consider constant nausea a necessary accompaniment, even in far advanced tuberculosis. There were dozens, perhaps hundreds, of remedies for upset stomachs on the market. Grace brought a dozen at least to the san. Finally, probably by chance—or fate—she hit upon a preparation that quieted my nausea so that I was able at least to eat; if not with pleasure, at least without fear.

Except for bed rest, fresh air, and plenty of food,

pneumothorax was the only method commonly used in the 1930s for the arrest of tuberculosis. This consisted of the collapse of the "bad" lung by the injection of oxygen into the pleural cavity. The bad lung, collapsed, was like a broken bone in a cast. It could no longer move, and in the lung at rest the lesions would thus have a chance to heal. The pleural cavity, the space between the lung and the wall of the chest, was filled with oxygen by means of a large (it seemed as large when it went in as a soda-fountain straw) needle, which was inserted between two ribs. Occasionally, the metal tube hit a rib. The sound that resulted was far worse than anything a dentist produces with his drill. A tooth is only a tooth; but your chest is where you live, heart and soul.

I wasn't eligible for pneumothorax because I had two bad lungs. If one were collapsed, the other would have to do all of the work of breathing, and this effort might bring on a fatal hemorrhage; or it might simply so accelerate the disease in the one lung as to make recovery impossible.

Since I couldn't have the real thing, I had a substitute pneumothorax: a ten-pound bag of shot, a foot long and six inches wide, was laid across my upper chest. With that in place, I breathed with the stomach muscles or not at all. The lungs were not subjected to any dramatic ups and downs.

Grace never again found anything as diverting as the cow that mooed. But she never failed to come and she always came laden. One day she brought (I kept a journal) the following:

1. Zinnias, red, yellow, orange-salmon
2. Roses, pink and red
3. A frosted root beer

4. A box of chocolate peppermints
5. A box of Cracker Jacks
6. A pencil sharpener
7. Ten new books
8. Two blotters
9. Beautiful purple grapes
10. A hyacinth-colored bed jacket.

When a year had gone by without improvement, it was decided to risk a pneumothorax. Because of old adhesions caused by pleurisy, not enough oxygen could be forced into the pleural cavity to collapse the lung. It was tried—again and again; but the tough tissue would not part. As a result of the efforts, no pneumothorax, but empyema, a tubercular infection of the pleura, resulted. Then the big needle went in to draw off the pus; this it *could* do.

At the end of the second year, they said to Grace, "Take her home and let her die amongst her loved ones."

Though I had a home of my own and a husband, they were five hundred miles distant. Max, my husband, was fully occupied earning the money that paid for my sanatorium expenses. And, in any case, he had not Grace's experience with caring for the sick or her skill. And no time at all.

I think that like many men who find themselves suddenly married to invalid wives, the ailing wife seems almost a stranger to him. How should he write her? Will accounts of his own health, of the activities she once shared with him, but from which she is now excluded, depress her? Writing was thus difficult. Visits, because of the nature of his work, were almost impossible.

So Grace took me home, but dying was the last thing she had in mind for any of her loved ones. It was not the last thing *this* loved one had in mind. If I had not been taken home, I had a plan that would put an end to endless night sweats and morning tears and afternoon burnings. I had become as impatient as the nurses with the tedium of tubercular dying. There was no very handy way for a bedfast patient to commit suicide in a sanatorium. Drugs were kept out of our reach. Guns and sharp knives were not available. I hadn't the courage or strength to fashion a hangman's rope out of bed sheets, then attach it to a door jamb, mount a chair, and exchange bed rest for something more everlasting.

There was, I knew (I had had "bathroom privileges" for a time at the close of the first year), a portable electric heater in the bathroom. I would have to crawl and I would have to choose a suitable time for the job, but I *could* get to the bathroom in this way; and there I could fill the tub, turn on the heater, get into the tub (this would be arduous), then lift the heater into the tub with me. Finally, bango! Electrocution! Sing Sing in the Sierra Madre foothills. No more T.B. for J. W. M.

I never (obviously) did it. But the possibility was a great comfort to me. If matters got beyond endurance, I need no longer endure.

Grace knew nothing of my electrocution by electric heater idea; and going home with her didn't seem to be a good time to bring it up. I was carried to the back seat of the old Packard by my twenty-year-old brother, then up the stairs at my parents' home to the room that was to be mine.

Grace and Eldo lived on an orange grove in one of

those houses called in Southern California "Swiss chalets"; houses with a smaller cube of bedrooms placed on top of a larger cube of rooms designed for other purposes. No better bedroom for a tubercular could have been designed.

Three walls were filled by windows. To the east was the canyon of the Santa Ana River and Old Saddle Back Mountain. To the south—my bed looked south—was the Pacific Ocean, Santa Catalina Island (visible on clear days), and the derricks of the oil wells on Signal Hill in Long Beach. To the west was the orange grove itself, the driveway to the house, and the palm trees that lined the driveway. In the palm trees lived doves and mockingbirds. A dove always sounded like a dove, but a mockingbird could sound like any bird, even that enemy of birds, a cat.

Through the palm trees and above the orange trees the Quaker Meetinghouse down the road (called Church in California) could be seen—and the singing heard when Meeting was held. A bell in memory of my great-grandmother, a Quaker minister, had been given to the church. But it could not be hung or rung until after the death of some of the older Friends who believed that bell-ringing smacked of merriment or papistry or worse. If such could be.

Beside my high hospital bed was a high, wide, stained-oak, mission-style table; commodious enough to hold a little radio, my eight-glass pitcher of water, a dozen books, ink, pen, journal, and the mooing cow.

A perfect room to live in: except that what I was doing wasn't exactly living. What I was doing, if Dr. Schultz was right, was dying. It was a good room to die in, too; and I longed to die and have it over with, and wept each morning to find that I was, alas, still alive.

Alive, but without any life to live. I had no future. I was completely without any spes phthisica by then. A house is not a home; a home is not a san. There were immediate changes and I was fearful of all change. The san had not rid me of tuberculosis, but it had rid me of the ability to live except on schedule. I was willing to die (I thought) quickly by electrocution. But I was not brave enough to face a normal, haphazard day. All Grace's days tended to be haphazard; no day was exactly like another, and none of them were like san days. Coming out of a san is a little like coming out of a prison. You have lost the ability to cope with a world organized for other purposes than looking after you.

Grace looked after me, all right; but she operated from the heart, not a schedule. Heart? What was that? At the san, we were mechanisms working on schedule or failing to work. We tried to keep heart out of it. Heart and emotion used up energy, accelerated respiration and sent the temperature up. All very bad for you. Going to a sanatorium is frequently necessary to survive; it can also be a way to learn not to live. I had barely survived when I came home to die among my loved ones; living I had completely given up.

With Grace, the living started immediately, and I seriously doubted I would survive it. But I had also learned at the san to keep my mouth shut. Complaints got you nowhere; and they also unnecessarily used up energy. I did not complain about Grace's heartfelt methods.

She took my temperature in an odd way. In the san, we lived and died by the temp stick. Its insertion and reading by a nurse was ritualistic. Life everlasting may be involved in the priest's wafer and sacramental wine. Life here and now was promised or denied us by the readings on the temp stick. At the same hour, four times a day, our temperatures were taken and recorded. If we asked, we were told what the temperature was. We learned not to ask. What we didn't know wouldn't hurt us.

Grace took my temperature, as she knew the nurses did, before breakfast. She went into the hall outside my room where the light was better, she said, to read the thermometer. She didn't wait to be asked what it was: she told me. T.B. morning temperatures are always, except in cases of acute illness, lower than afternoon temperatures. What she repeated was not bad: not so low as to suggest that I had died in the night without knowing it; or so high as to foretell a major flare-up.

In spite of all of the unorthodoxy of Grace's care, my temperature continued, tenths of a degree by tenths, to fall. Up occasionally, a tenth, but the general trend—down.

I became suspicious. What went on out there in the hall where the light was better?

"Could I see the temp stick?" I asked.

"It is called a thermometer."

"I know. I picked up the temp stick habit at the san."

"Of course you can see it. These bifocals sometimes play tricks on me."

It was what she said it was: occasionally a tenth or

two-tenths of a degree off. But even the nurses sometimes misread the exact point to which the red had climbed.

What went on out there in the hall? Did the thermometer sometimes get a good shaking? I decided to forget it. I could no longer add to my weight with jars of night creams and bottles of hand lotions on the once-a-week occasions when I was assisted to the bathroom scales by my bedside. If there was witchcraft in the hall, so be it. What did it matter? Why, destined to die, insist on each day's full quota of suffering?

"Why don't you turn on your radio?" Grace asked.

I couldn't. The radio brought to me the world outside my room. I didn't want to think of it, to be tormented with reports of people who had lives to live; who went to work, attended seminars, played tennis, swam at the beaches, bought food at the grocery store. I had a plea so modest I thought that it would endear me to fate: let me walk unblemished by any afternoon flush, or made noticeable by any hackings or rales into a grocery store and there buy a head of cabbage or a loaf of bread.

The past was equally unavailable to me, was even more heartbreaking. I had had so much: husband and home; studies that exhilarated; a body that could climb hills and aquaplane and lie in the sun; all impossible or forbidden now. And I had taken them all for granted—everybody's gift, I thought then, and for time unending.

I could not live in either the past which *was* past, or the present from which I was locked away. Where could I go, in my mind, except to play games with it: remember the cast of characters of novels read long ago, then assemble a cast of movie actors to play their parts. For David Copperfield's Dora, Doris Day? For *Vanity Fair's* Becky, Bette Davis? For *The Scarlet Letter,* how about an irresistible preacher like Humphrey Bogart? Movie-casting isn't exactly a life—it at least stops short of madness.

Grace knew, though no word of mine told her anything, the limbo in which I lived; somehow, without words, she knew. Since I had no life of my own, past or future, in which to live, she gave me her own life, as a young woman, as Grace Milhous, a Quaker girl on a farm in southern Indiana at the turn of the century.

I was still not strong enough to follow a book, line by line, paragraph by paragraph, to its conclusion. But I could listen to short stories, and anecdotes, some hilarious, and all acted out by their narrator, a former elocution prize-winner. I was capable of reading a word at a time. I could look at a dialect dictionary; I was strong enough for six words whose continuity was nil, checking words Grace remembered against those listed in the dictionary. I wrote them in my journal, where the lists formerly had been of temperatures, weights, letters received from my husband. I was moving a little outside my self, my physically functioning or nonfunctioning self.

When she was a girl in southern Indiana, Grace had used such words as these:

1. Work-brickel, meaning energetic
2. Clever—hospitable
3. Smidgin—scant amount
4. Fine-haired—priggish and snobbish
5. Dauncey—nauseated

6. Feather into—start a fight
7. Fractious—especially of horses—nervous and easily upset
8. Middling—halfway between bad and good
9. Sorry—poor or inferior
10. Zany—strange or queer
11. Infare—party, the second night after the wedding, usually held at the groom's home
12. Playparty—a party of singing and stylized bounding about which was a substitute for dancing
13. Woodscolt—an illegitimate child
14. Hockey—manure, human or animal

These were words Grace still used, though as her California college-educated children grew up her vocabulary changed.

Little by little, living with Grace, like the frog kissed by the maiden, I became human again. Not the human I once had been; my own past and my own problematical future were still areas too painful for me to enter. But Grace's past I could enter and explore and live in. It was better than movie-casting; her story was real and she, the heroine, sat by my bedside with her own gestures and accents. And Grace's gift of a life was not like a Christmas present fabricated from duty, a tit for a tat, which the season demands. She enjoyed the making as much as I did the receiving; and what she made was engrossing as only "This is the way it was in that far country where I lived when young" can be to a child. For, short of diapering, I was cared for like a child in almost every physical way. The hospital bed was my crib. I was only beginning to be able to read again. Like an infant, I had the present or I had nothing.

Grace gave me southern Indiana in 1900. I emphasize "southern," because in its language and cooking and customs it was nearer Kentucky than Ohio or Illinois. There in the land of Grace's girlhood I lived, nearer to its climate, its topography, its flowers and trees, and to the farmhouses of my grandparents than to anything in my own state of California.

I have never known the misery or inconvenience of snow. From Grace I learned only of its silence and mystery. "The sky whispered," she said. "And yet it was a whisper you could see." Feathers, sometimes, curly and white, verging at times on sleet or hail. Sticky, sometimes; hard and dry at others so that the runners of the sleighs rasped like a plane cutting wood.

When wet snow clung to fence posts and outhouses and corncribs, it resculptured them then into shapes Oriental, or even extraplanetary. Snow was good to eat. With sugar and cream it was winter ice cream. It was blue in the shadow of pines, pink in a fiery sunset. You could jump into a snowbank and it received you like a feather bed, carefully, gently.

Perhaps Grace's snow had for me some of the appeal of Scott's Antarctica: it was cool. But only some of that appeal. Chiefly, it was a part of the only landscape I could live in comfortably. There was no pain there for me. It was nothing I once possessed and had lost; it was not a future forbidden to me.

Small streams where Grace had lived when young had peculiar names: they were called "branches"; and larger streams were "creeks," not "rivers"; and "creeks" was pronounced "cricks," not "creeks."

Oh, it was a strange, haunting land where Grace grew up; nothing like plain, humdrum Southern

California. Ordinary words were pronounced in unusual ways. Florida was "Floridy"; peony, "piney"; and California itself was "Californy." Tuberculosis was "consumption" or "lung fever." Cooked breakfast food was "mush," and there was never a meal without gravy.

Edible foods grew wild. In Southern California, after you had eaten the cactus apple (which was too well protected with spines to be very inviting) and elderberries, you were about through with what was wild and edible.

Where Grace grew up you could munch your way across the landscape. In the fall, there were pawpaws, which tasted, Grace said, like bananas but juicier; wild persimmons, after the first frost, were smooth as butter and sweeter. The meat of a hickory nut was so well protected in the nut's tiny compartments that it took an afternoon's pounding with a flatiron and tack hammer to get your fill. Butternuts were less trouble, but less tasty, too. Fox grapes, small and with leathery skins, grew along the creek banks. Wild strawberries were almost too pretty to eat. In the springtime, Grace, hungry for green food, climbed oak trees and ate leaves the size of mouse's ears well sprinkled with salt.

Thunderstorms, which California lacks except in the mountains, were accompanied by a variety of forms of lightning: ball, chain, heat, and sheet; all left in the air a smell of scorched feathers. The way to escape harm from lightning was to get into a feather bed.

There were cyclones and twisters where Grace grew up. They appeared on the horizon like elongated tops. When one was sighted, "Head for the springhouse or root cellar" was the order of the day.

There were buildings on the farm where Grace lived never heard of in California. Carriage houses, corncribs, haystacks, smokehouses, privies, cowsheds, farrowing pens, springhouses.

The springhouse was half-cellar, half-house. You went down steps into a room partly underground. Along each side were wide stone troughs through which spring water to the depth of three or four inches ran. In these troughs, kept cool underground, pans of milk, crocks of sausage meat put down for gravy "timber," jugs of cream, some sweet, some ripening for churning, were kept. There was piccalilli there, and apple butter and cider, sweet, of course. Grace's father would not have hard cider on the place, let alone applejack, which drove hired men crazy, and even got to an occasional hired girl—and worst of all, occasionally, to both at the same time.

In those days, a little craziness didn't put you in the lunatic asylum. It was taken for granted; lived with. Grace went with her mother to take food to a sick neighbor lady. In bed with the lady in the middle of a fine afternoon would be her husband, suffering from no ailment except sympathy. Sympathy, he said, always overcame him when one of his wife's ailments put her to bed. At such times, he was good for nothing in field or barnyard. He was forced to go to bed with his wife in order to share her suffering. The neighbors, knowing that when Mary Ann was bedfast, Henry would be also, took enough invalid food for both.

They had their babies together. This was an unhandy arrangement for the neighbor ladies who were the amateur midwives. They needed all the bed space for the job of delivery; but Henry did make himself as small as possible, and did lie on the far bedrail.

Once the delivery was over he got up, good as new, and went right back to hoeing corn or milking the cows. Henry was a fine man in every other way; and though it could be unhandy, there was nothing evil about an overendowment of sympathy for his own wife.

There was a seamstress in the neighborhood, very good, but who would sew only for men. Women, she said, used all their clothing, from kitchen aprons, shimmies, drawers, and corset covers, right down to sunbonnets and mittens, to lead men on. She would have no part in this hanky-panky. Men, insofar as clothing went, were pure. All they wanted was to be warmly and conveniently covered. Even a man's nightshirt was designed for his comfort while sleeping, not to seduce his wife; or worse still, some other man's wife, with a come-hither garment of outing flannel.

This woman, whose name was Etta Nisewander, followed her conscience and had all the sewing she could take care of sewing for men only. There were some catty women, of course, who said that the clothes Etta made for herself weren't as plain as what she preached for others. But Grace's mother sent Grace's father, a sober Quaker, not likely to have his head turned by a lace jabot or a bouncy bustle, to Etta for his overcoats. Her own clothing and her children's, Grace's mother made herself.

While I lay in bed, with no life of my own to look

forward to, or that I could endure to look back on, I was given this other endurable life to live in, to wonder over, to speculate about. Not yet to laugh at, though there was much Grace said that was funny. I was afraid to laugh. It would be tempting fate. It would be whistling before I was out of the woods. Grace expected me to get well, and talked to me about the time "when you are well." "Even if I should get well," I told her, "I know I will never be able to smile again."

Grace, though this was laughable, didn't laugh. If that was the way my face felt, all the logic in the world couldn't contradict my muscles.

If the life Grace gave me to replace the one I no longer possessed had not been reported to me so vividly, or if it had not been the life of people of my own blood (my life before I was born, so to speak), I could not have lived in it as I did.

I knew the floor plans of her home and of her grandfather's home. I knew where the grafting table (Grace's father and grandfather were both nurserymen) stood in the kitchen at her father's; I knew where the homemade phone was placed on the stair landing at her grandfather's. I could have walked into either house in the dead of night and found the spare bedroom in pitch-darkness.

I knew the names of the dogs and horses, the hired men and the hired girls. Grace had a pet heifer whose name was Elizabeth. The bird in the cage, though it was a starling, not a parrot, was named Polly. There were no cats. Grace's mother said that cats were sly and dirty and Grace herself inherited this belief.

There was a rug in front of the organ in the parlor.

Woven into the texture of this rug was the likeness of a dog, a Saint Bernard, one would guess. This dog, like the real dog in the yard, though they looked nothing alike, was also called Old Pedro.

So, though I was bedfast insofar as my body was concerned, Grace provided me with a new country in which my mind could travel; with strange people, some of them blood kin, I could think about.

There was an uncle who could not resist buying oddments, the older the better. He was drawn to what had once been used, loved, and later abandoned. He had a dear wife who could not endure clutter. To indulge both his loves, Uncle built a series of sheds to house his purchases. He had a set schedule for visiting these sheds; much like a Mormon's schedule for visiting his wives. Uncle loved all his possessions and labored to see that none should feel neglected.

There was an aunt (she might have been the wife of this uncle, but wasn't) who was so tidy that she kept the tree stumps in her yard shining, not only swept and dusted but scrubbed.

Grace had had her own sorrows and surprises and dreams. On the way to school, a young man she admired looked down from the horse he was riding and said, "Good morning, sweetheart."

Grace, who wished she was, but knew that ten was too young for that sort of thing and who did not want to be thought tongue-tied as well as immature, answered, "I am not your sweetheart."

"I never said you were," replied the dashing young horseman, humiliating Grace with this proof of her forwardness.

Grace and her father were birthright Quakers,

members whose forebears had been "convinced" in Ireland by followers of William Penn himself; and who had come to America on the ship that brought Penn. The Meeting Grace attended was silent, unlike the California Meeting I knew, which was brisk and noisy, with a hired preacher whose sermons were as pulpit-thumping as anything a Baptist or Methodist could deliver.

But though there were no sermons, the Meetings Grace attended were not completely silent. Those in the congregation who felt a concern to share some holy insight, some prayer, some Bible verse with their fellow worshippers might rise and do so. At the age of eight, Grace rose in Meeting for the first and only time in her life, and shared with the congregation a concern. It had to do with her brother, Walter, two years older than she.

Walter had convinced her that, to be on the safe side, since Jesus Himself had been baptized, she should, in spite of Quaker practices, be sprinkled. He himself would do the job. This at least was good Quaker theology (insofar as Quakers can be said to have any), since George Fox declared, and Quakers believed, that each man, having that of God in him, was a qualified minister of the gospel.

Walter would, he told his sister, take her down to the branch and there in the name of God sprinkle her. Grace could not see that this would harm anyone; and it would obviously give Walter a good deal of pleasure. (Walter was born in the wrong faith and the wrong area to make use of his dramatic gifts as either a Bible-thumping evangelist or a stand-up comic.)

At the branch, Walter sprinkled Grace as he had

promised, saying as he did so, "I baptize thee in the name of the Father, the Son, and the Holy Ghost." The ritual was certainly un-Quakerish; but it was downright heathenish for Walter to do his sprinkling as he did by peeing on his sister.

Grace did not tell her parents. She was no tattletale, and the tanning Walter would get would not erase the humiliation of what had happened. Peed on! Not even animals did that. So she took it to the Lord in prayer, and out of that prayer came the concern which she shared with the Meeting.

First, she read scripture to them. Mark 11:24. "Therefore I say unto you what things so ever you desire, when ye pray believe that ye receive them and ye shall have them."

Next, John 14:14. "If ye ask anything in my name, I will do it."

Having read those verses, Grace prayed, "In Thy name I ask that my brother be forgiven for and cured of his abominable practices. Amen."

Peeing on his sister was not the abominable practice (though it was that) the members of Goose Creek Meeting had in mind. The Meeting, except for a concluding prayer from one of the recorded ministers, *was* silent after that. Completely.

Grace's father and mother could not ask Grace what she meant by "abominable practice" without proceeding with some sex education they were not prepared to give her. Walter, smart stand-up comic at ten, said, "It isn't a practice. It was my first baptism."

Baptism, whatever the method, contradicted the beliefs of the Religious Society of Friends.

Walter went to the woodshed with his father and

his father took his razor strop. What went on out there was so prolonged and noisy that Grace prayed her Meeting prayer all over again, but with a different request. "Let Papa stop whipping Walter."

That night, after he was in bed, Grace went to her brother's room to ask forgiveness.

"I am sorry about what I said in Meeting, Walt. Now you can baptize me with branch water and we'll both be cleansed of our sins."

"Too late," said Walter. "The baptism you got was of the devil. You are his own now. Nothing can change that. You are his child."

"Aren't you the devil's child then? You did it."

"When I did it, I was. But I was baptized with fire in the woodshed and now I belong to Jesus."

For a long time Grace tried to persuade her father to whip her just as hard as he had Walter so that she would no longer be the devil's child.

"Did that boy tell you that you were the devil's child?"

Grace saw what would happen if she said yes would be another trip to the woodhouse for Walter. She kept quiet; but Walter's threat haunted her for a long time.

So Grace gave me the world she had known and the life she had lived and the dreams she had dreamed for my own use. But she didn't for a minute believe that they should be a substitute for my own world and my own living. And to have those I would have to regain a body that could stand and walk again, even run and jump.

"The orange juice here is so much more yellow than the orange juice we got at the san," I told Grace.

"What's the use of living in the middle of an orange grove if you can't have thick, yellow orange juice?"

I didn't know the answer to that. I didn't know either that even in the middle of a parking lot you can have thick, yellow orange juice if you add two eggs to each glass of juice.

I knew nothing about the eating habits into which I had fallen at the san.

"Jessamyn," Grace said, "do you have to eat like that?"

"Like what?"

"Every bit of one thing until you've finished it, then every bit of the next dish."

"Do I do that?"

"All the time. All the potatoes. Then all of the meat. Next, all of the salad. Wouldn't it taste better if you had first a bite of this, then a bite of that? That's the way most people eat."

I had forgotten it. At the san, all foods tasted unappetizing, whether because it really was so or because I was sick, I don't know. But I went there determined to get well. Since eating was one of the recommended ways, I went at eating methodically. If supper was a boiled potato, a dead sardine, a sliced tomato, a glass of milk, and a dish of tapioca pudding, I put away sardine, potato, tomato, tapioca in orderly fashion, one at a time. The milk was last. It washed some of the taste away.

Grace's menus contained no dead sardines, no boiled potatoes, their eyes still glaring at you. Her food did not need to be downed by exercise of will power. But it was difficult to relearn the association of food and pleasure. Food at the san was a necessary evil. There was more pleasure in the sense of power obtained from eating bad food because it was good for you than in eating the tasty food Grace provided because it pleased your taste buds.

It is a mix-up of the same order that causes some persons to find sex as pain more satisfying than sex as pleasure. I do not yet understand why I felt it weak to begin to enjoy food. Was it pride? *Anyone* could eat what tasted good? I had trained myself to eat whatever was put before me, even when a cockroach, as was occasionally the case at the san, shared my tray. Physically, I was perhaps getting stronger. But what was happening to my moral fiber? This was the Calvinism of the sickroom. The egotism of the invalid. I am better than you because I suffer more.

True, I was losing some of my taste for suffering. I was learning to eat, not because I was strong-willed, but because food was becoming appealing. I lost more of my taste for suffering when Samantha arrived.

Samantha was a six-week-old kitten brought to the house in a shoebox and offered to Grace as a bedside solace for her dying daughter. Grace disliked all cats; and a cat in a sickroom seemed to her especially ill placed.

I myself had never had a pet of any kind: bird or dog, cat or horse. The oldest of four, with an ailing mother, married young, a would-be scholar, where was there time for *animals?* Now there was time. Now there was time—or never.

I know what I felt when I put my hand on that ball of living fur. Grace saw what I felt. But a cat's litter box in a room already furnished with a bedpan?

Thank God that wouldn't be necessary. The casement windows of the room were always open; and they opened out onto the lower roof. By propping one screen a little open, the kitten could come and go as it wanted and, one hoped, as it needed.

Grace said, "Thank you. It will be company for Jessamyn."

Not only company: sanity, health. Not right away, of course. Cats have magic, but in stubborn cases it takes a little time for their magic to make itself felt.

Dr. Wehrly, who visited me three times a week, was like Father Flannagan at the san, a plump ex-tubercular. He said, "You began to get well the day that cat arrived." He said this a year after Samantha's arrival. Samantha provided no overnight cure, though she did provide immediate interest in something outside my own physical functioning.

There had come to be nothing else. I was my own entire world. My sun rose and set with the mercury on my temp stick. The rales in my chest were all of nature's sounds I heard: surf of sea and sough of wind in trees and bird song outside the windows.

I had developed fantasies to ease my pain. The attempts that had been made to tear down the adhesions so that I could have a pneumothorax had left me with a burning, aching pleural lining.

It was hideous to believe that part of my own body was treasonably working for the painful downfall of my life. I needed something to fight—and not myself. I was against a personal civil war. So when the pleural pains were worst, I imagined: rats! Rats were gnawing the flesh around my ribs. I could fight rats, not give in to rats, never by any sound give rats the satisfaction of believing that they were winning. At night, when the cutting pain lanced across my chest, I said, "The rats have come," and though this did not stop the pain, it absolved me from the need of fighting myself. I fought the rats instead.

When misery was at its greatest, I left my room in fantasy (was Oates my example?) and lay in cold weather, on hard clods, in the middle of the orange grove.

It was a comfort to have the source of my discomfort outside myself. I lay there (in my mind) and thought that I would never again move, would wheel round and round, a part of the orbiting earth and the circling stars forever—hurt perhaps, by the pelt of winter rains and the sting of sand driven off the desert by howling Santa Anas, but not self-hurt. Just a part of the general ache of being.

There was no one for whom *I* could do anything. All was done for me. My face washed, my body bathed, my toenails cut (yes, even that), my bedpans emptied. I awakened each morning crying to find that I was still alive.

I stopped that (the tears still came, but they ran unobserved down the back of my throat) after Grace said one morning, "You say you can't do anything for anybody. You could lighten my whole day, petty, if I could come up here some morning and find that you had stopped crying.

"You are going to get well," she went on. "If you weren't, I would put that electric heater you told me about in the bathtub and get Merle to carry you downstairs and put you in the tub with it."

"I never told you about any electric heater."

"You may not have known what you were saying, but you told me."

"Don't you think that would be wrong?"

"If you were going to die anyway and all the life left to you was suffering or being so full of dope you didn't know who you were, I think quick dying would be better than slow."

"Did what happened to Grandma make you think that?"

"It helped. If I had it to do over again, I would have found some way to save Mama those last two months of agony."

I knew something about those last two months. I was nine years old at that time, and I remembered my grandmother's screams as she lay dying of cancer.

"I would do it for you. But you are going to get well. And you can do something for me. Don't cry every morning."

So I swallowed my tears for Grace's sake, but there was no more pleasure in their taste than there had been in san food. The pleasure was in will power.

Samantha was pure pleasure. Part of the pleasure, I'm sure, came because she needed me. She was homeless, except for my bed; motherless, except for me; hungry unless I fed her. But I fell in love with her. A guinea pig or a hamster might have needed the same amount of care and failed to divert me from my concern with my own needs.

You fall in love with animals as you fall in love with people. A computer could not possibly analyze the elements involved. Perhaps you fall in love when you need to fall in love. At that moment in my life I might have fallen in love with a prepossessing chicken, or even a spider like Charlotte. Love for Samantha, I think, would have been pretty irresistible at any stage in my life.

She was what Americans call a calico in color; what the British with greater elegance call tortoise-shell. She was long-haired, a mixture of browns, oranges, and blacks; she had white feet, a white belly, a bib of white, and white whiskers. Her eyes were round and golden. She was young and home-less, an orphan dependent on handouts. None of it showed in her bearing. She walked back and forth under my chin to give herself the sense of being mothered. Her purr, though she was staunchly built, round and fat, shook her the way the engine shook old Model T Fords.

Whatever I ate, she ate. Milk, of course, which I always had; spinach, tamale pie, fried eggs. She would have sipped the revved-up orange juice, too, if I had given her a chance.

Grace, in spite of her dislike of cats, developed a shamefaced tolerance for Samantha. The kitten was smart, and Grace was always a sucker for smartness. Once Samantha was put out the window and onto the roof, she accepted the sanded tar-paper roof as her private comfort station and made the jump there-after unassisted through the always open window and onto the roof.

I did more than tolerate Samantha. As I said, I fell in love with her. Sickness had robbed me of a taste for food and I knew it. I knew that I could not stand or walk; and that a paragraph of reading was all I had strength for. But I did not know how much I had missed touch and touching. Samantha craved touch. She asked for and bestowed caresses. She would walk over my shot bag, reach my face, and there press her face to my lips. She surely was not bestowing feline kisses; but kissing was what it looked like. Since I could tell when the trip over the shot bag started, and what was impending, I would say, "Give me a kiss, Samantha," and so impress visitors with my trained cat; and with me, too, a Clyde Beatty of the sickroom.

Samantha slept under the covers at my feet. How either of us survived this arrangement, I don't know. Samantha survived, perhaps, because I was often wakeful, and on these occasions she had a chance to come out from under the covers, play games with me, and get a breath of fresh air.

The game we played most of the time was cat-and-mouse: Samantha was the cat and my fingers were the mice. Behind a barricade of bedclothes, my fingers twitched like nervous mice: on the other side of the ridge, alert eyes peeping over so that the mice would not escape, crouched Samantha. She pounced, the mice ran for cover. Then Samantha, as if she knew the game was a game nearly as well as I, repositioned herself behind the mound of quilts to wait for more mice; mice who always reappeared.

Playing cat-and-mouse with Samantha, I forgot about the rats; nor did I want to be on a bed of clods in the middle of the orange grove. Clods would not provide a suitable bed for a kitten.

Half-grown, Samantha began a practice I have never known in another cat. (And since Samantha I have never again been catless.) She carried up the stairs to my bed objects that it amused her to carry, or which she thought might amuse me: a spoon, a glove, the stopper to the downstairs washbasin, a kid curler; and once a mouse dazed, but not dead. She seemed to have no intention of eating it, and in time the three of us might have learned to live together. But cats were as far as Grace could bring herself in acceptance of animals in her household. The mouse and the other donations Samantha made during the day were removed each evening to places Grace thought more suitable for them than my bed.

I could talk to Samantha. She did not know about the past that I had lost; nor about the future that I dared not believe in. We talked about the present, which was all either of us had.

The nights of talk and games increased as sleep lessened. I knew that sleep was a part of getting well, and for this reason, I willed myself to sleep as I had willed myself to eat. But you cannot will yourself to sleep. The unconscious is in charge of sleep, and the unconscious has a will of its own. Coerced by conscious will, sleep becomes recalcitrant; order sleep and it leaves forever.

I feared the nights. Talk to Samantha and games of cat-and-mouse were not enough to fill eight hours after they had been all the sixteen previous hours had held. I tried not to hear words that meant that night was coming. Beautiful words like twilight, sundown, moonrise were ugly to me. They said that bedtime was near; and bedtime was intended for sleep and I could not sleep. If I wanted to live, and I did, it seems odd that I so feared and resented sleeplessness: for surely you are more alive awake than asleep.

But I was caught in a bind: to recover, to be truly alive, sleep, I believed, was necessary. I could not manage it. I would die wide-awake. In my room of windows, I saw the moon set and the sun rise. I began to long once again for the clods.

One night in the gray of fading stars and before the sun had pinked the sky above Old Saddle Back, I knew, without turning my head, that I was being watched—and listened to—for I was talking cat talk to Samantha, who was too sound asleep to even purr in response.

Grace came into my room. My door was always open between Grace's and Eldo's room, which was down the hall from mine, with only a bathroom in between.

"You haven't been asleep all night, have you?"

"It doesn't seem like it."

"This is the third time I've been here. You haven't."

Grace stood by my bed, in her long-sleeved, flower-sprigged outing-flannel nightgown, nobody's picture of a plump, gray-haired mother, but the picture of a woman who cared.

"This can't go on."

"How do you stop it?"

"Medicine."

"You mean sleeping pills?"

"Pills or liquid or plasters. I don't care what."

"They wouldn't permit it at the san. They said no use curing us just to have us turn out dope addicts."

"This isn't the san. Tomorrow night you'll sleep. The pills will put you to sleep. And God will not let you become a dope fiend."

Grace, after her experience with Walter, never again voiced her prayers in Meeting, but she was a strong, perhaps violent, pray-er; a woman God couldn't easily ignore. The combination of pills for sleep, and Grace and God to counteract the pills' addictiveness, gave me confidence at once. The sun was way above Old Saddle Back before I awakened next morning.

The pills were not pills but a thick brown broth; it was made from what appeared to be bouillon cubes. This sleeping potion, made in Switzerland, I think, was called Sedobrol. It may have been as useless as real bouillon cubes for sleep, but the thick dark broth tasted like sleep; even tasted like, I imagined, that potion Romeo drank in the crypt. I had more fear of never waking up than of not sleeping at all after I downed it.

Full of Sedobrol, and with my cat at my feet, I slept like a Pharaoh. My sleepless nights were over.

I still do not understand the torment of sleeplessness. I prefer pain—which also prevents sleep—to causeless sleeplessness. Sleeplessness caused by pain is understandable. It gives you something to think about, to analyze, and to curse; you can despair. You are not alone in a vacuum. Your pain is there with you. You can howl.

But sleeplessness without pain is like death without dying. You are suspended in space; the world you knew far behind you, no other in sight. Sleeplessness is a desert without vegetation or inhabitants. Only you, under a curse of some sort, alone; not mindless, but with no use for your mind; without pain, but with a body that needs pain to remind it that it still lives.

Grace, after that, soon began to find the light in my room strong enough for her to read the thermometer there. (Actually, it had always been brighter there than in the hall.) And she would hand the thermometer to me to read saying, "See if you see what I see." I saw what she saw. I saw that my temperature was going down.

Without any ballast of bars of soap or bottles of ink, I saw that my weight was going up.

Eating, sleeping, getting fatter and cooler, I was still afraid to re-enter the world I had lost. I had learned to live on little. Why risk the possible disappointment of learning to want more?

Grace was not a meager liver. One evening she said, "Turn on your radio for one minute," a thing I had never yet done; for out there, when the knobs were turned, was the lost world where the healthy lived and which would pour in upon me, still chained to my bed.

"One minute only, and you can turn it off."

For Grace's sake, for I was not too self-centered to be blind to the fact that if I was chained to my bed, she was chained to me, I turned the knob. What I heard was Bing Crosby singing "Wagon Wheels."

Neither the song nor the voice threatened me. Too much rhythm set my stomach to swinging. Too much beauty or emotion, Mozart or Beethoven, made my heart thump and my face burn. Der Bingle, who had enchanted thousands, many no doubt in worse case than I, opened a small-sized door for me into the world I had lost. I have never turned off the knob I turned on that evening.

Next came, paragraph at a time, page at a time, and finally book at a time, books. I had lost much of my graduate-student snobbishness. I would take a look into books never mentioned by J. S. P. Tatlock or T. K. Whipple. Some of the books that turned up in Grace's traveling-salesman book-carrier case would have had their approval: Giles's *History of Chinese Literature,* for instance.

"I have always wondered what kind of stories the Chinese wrote," Grace said. "That book ought to tell you." It did.

She brought home *A Beginner's Book of the Stars.* "You've got practically an observatory here. By the time you leave, you could know all their names."

By the time I left, I was familiar with the sky to the south, the east, and the west. What went on overhead or to the north, both invisible from my bed, was something I had to take the *Beginner's* word for.

As my ability to read increased, I went on to subjects more exciting, though not of more import than the stars or Chinese literature. It was the time of the first trumpetings of new talents: Hemingway and Faulkner and Wolfe and Eliot and Fitzgerald. The librarians hated to see Grace enter their door with her enormous list of books to take out or reserve. So long as she stuck to the Chinese and the stars there was not much competition; but everyone wanted to read *Look Homeward, Angel,* and *Tender Is the Night.*

As for me, I had relearned to read, but I had not learned to read calmly. Characters in books were almost the only people I had; and they are always ten times as exciting as the people one usually meets—in a sickroom, at least. A book was my party, my rally, my Fourth of July parade, my theater, my sermon by Billy Sunday, my U.C.-Stanford big game.

The books had to be put outside my room at night so that their words wouldn't reach out through their covers and set my nerves atingle.

Grace, who endured the presence of cats, arranged for the absence of books; books that she had wrangled from balky librarians and transported up steep stairs, she removed at bedtime.

After a year, the day came when Dr. Wehrly decided that even without thermometer misreadings in the hall or my own scale management by means of bottles and bars, I was well enough to make the trip to his office, twenty miles distant, for an X-ray. The proof of what was going on in the lungs lay in the lungs themselves—and only an X-ray would tell that story.

Grace bought me for that momentous occasion a pair of what were then called lounging pajamas, a very fancy pair, wine-colored, with gold buttons and frogs. A dress (pants, not then used except for hiking) was too drastic a change after years of pajamas. I would not know how to manage skirts any more than did the masquerading Ulysses.

The lounging pajamas, more ornate than I would have chosen myself, I donned. Then Grace brought out a hat to complete the costume, pink horsehair with flowers. A hat was bad enough in itself; but pink horsehair, white Shasta daisies, *and* gold-frogged lounging pajamas would, to my mind, give me a setback that might show up on my X-ray. I was by no means yet an arrested case. An outfit like that would be more than whiting the sepulcher; it would be gilding it.

I took that hat and kicked it under the bed. Samantha jumped out of the window. Grace sat down in the rocking chair and laughed and laughed and rocked and rocked.

Later she told me, "The minute you kicked that hat under the bed I knew the battle was won. Up to that time you had been meek as Moses. No spunk at all. Eat what was put before you. Content to make a cat your whole world. Not caring what you looked like. I tell you, nothing ever gave me so much pleasure as the kick you kicked that afternoon. You were your old self again."

"Kicking hats? That was my old self?"

"More so than cat talk and making yourself think that you could get well by loading up your bathrobe pockets before you were weighed."

The X-ray was good beyond all expectation. In four months I went to my own home—again.

I had gone there at the end of my first year in the sanatorium. I believing, as did my husband, that one recovered from T.B. as one did from measles: that there were no recurrences.

In six weeks I was back in the sanatorium in worse state than I had been when first I arrived there. I did not come walking this time, but by stretcher and gurney. I came back heartbroken; not so much at my own relapse, which of course did not make me happy, but with what my re-entry meant to my fellow lungers. I had done well that first year: no more hemorrhages, lesions lessening, fibrosis increasing. My departure, in dress and hat then, with the blessing of the staff, was to every patient a promise of what he could do. I was a female Father Flannagan, whose sickness and recovery they had seen with their own eyes. I was hope.

Wheeled to my bed, no bathroom or dining-room privileges now, they saw their own fate. I was despair. I was what awaited them. I was hope dashed: a final hemorrhage or a "let her go home and die among her loved ones." I had committed an act of treason to those closest to me. I was the lookout who went to sleep; the man in the trenches who failed to go over the top with his comrades. I was sorry for myself, all right. Nor was I ready to die for my fellow prisoners; but I was fully aware of what the manner of my return meant to them. It meant, "This is your future."

When I went to my own home the second time, there were certain likenesses to the first: Grace, as she always did, waved until the car I rode in was out of sight. But this time I went by Pullman, not day coach. This time Samantha accompanied me in the baggage coach. This time there was no clamorous welcome-home party, but a bed with covers turned back awaiting me and a housekeeper to unpack my bags.

No longer did I live like one who has been ill and is now recovered. I lived like an invalid. Kicking the hat under the bed may have convinced Grace that I was "my old self." But "my old self," somewhat livened up while Grace was there to liven me, was a person who had known a terminal ward, and heard the hearse climb the hill at 3:00 A.M., and experienced the return to the san after only six weeks of "normal" living. She now believed that half a life was better than none.

My life was lived between bed and sofa. I parted with my shot bag, but only at my doctor's insistence. I took rest periods as in the san, and kept records of my temperature and pulse rate. I no longer put weights in my bathrobe pockets—and I no longer needed to; my weight had gone up from 120 to 170. My stomach had begun to function properly again, and I gave it plenty of work: malted milks and beer (which I had learned on doctor's orders to drink) and creamed chipped beef.

There was one life, in spite of my self-imposed restrictions, I could live. To use a clipboard, pen, and paper did not require much physical effort. They could be used in bed. To write had always been my dream. A doctorate in English literature was only a diffident would-be writer's way of staying close to words without taking the writer's risk of exposing his own perhaps pretentious ineptitude in the use of words.

In a recent book, *Conundrum,* Jan Morris says that she (then he) at the age of three or four knew that though he had a male body, he was a female. At the age of three or four in southern Indiana, children had not yet heard, let alone known, the meaning of the word "sex." But at that age I knew that I was born to read and write, and sat in a corner crying because I could not make out the words in the book I was holding. Learning to read was not a painful process. But I came to writing in a manner as prolonged and painful as James Morris's in coming to womanhood. Writing, he has managed apparently without pain in both genders.

I am not proud of the manner in which I came to writing; unwilling, until I was backed into a corner by disease, and unable to do anything else except perhaps crochet, to pick up my pen. Talent is helpful in writing, but guts are absolutely necessary.

Without the guts to try, the talent may never be discovered; and its possessor, virgin to ink and words, may, like any other castrated person, exchange potency for sourness.

It was not pretentious for a semi-invalid to use her pen. It was reassuring to onlookers to see her so cheerfully engrossed in one of the few activities left to her. There was less call for pity. (Though writing *and* reading to nonwriters and nonreaders *does* often seem rather pitiful. Wouldn't they rather *live?*)

I had my pen and my clipboard and it did not take guts to use *them.* Writing had become laudable. But what was I to write about? Certainly nothing that had anything to do with the past years of my life. I could not even see the words "the consumption of beef was at an all-time high in the United States last year" without experiencing a rise in temperature at the sight of the word "consumption."

But I had the life Grace had given me. It was not her own life. It was not even the life of her parents. It was the life of which she had only hints and inklings, but about which she had wondered and dreamed, of her grandparents. She had not told me anecdotes about them. An anecdote is as self-contained as a pebble. It's a finished product, useful if cobblestone is the product you desire, and all you need to supply is cement.

What Grace gave me would not work up into cobblestone. It was too tenuous, too shimmering, too hearsay. There was a landscape there; that was solid. She knew the smell of spring flowers and the colors of fall leaves and the sound of branches flooding at springtime with melted snow. She knew the spirit of Quakerism, of those Irish and Welsh Friends who had come here at the end of the seventeenth century and the beginning of the eighteenth—some with Penn himself. She had letters written before the Revolution. What were they like, the men and women who wrote these letters? Who sailed out of Cork and Plymouth believing, "There is that of God in every man." She wondered. So did I.

But I had a pen and I had leisure. I could put my wonderings down on paper. I could give those distant, unknown Quakers names. I could put their "thee"s and "thy"s on paper; write about great-grandfather's hearsay love of music and fast horses; and make up out of whole cloth their unknown, to me, stance during the Civil War.

I did so. I wondered in words and gave the words story form.

Sickness had made it possible for me to write without appearing pretentious. It had not given me the backbone to send what I had written to magazines or publishers. To do so would be to assume that what I had written was amusing, or enlightening—or at least readable. It appeared to me to be a presumptuous act.

My husband persuaded me that the presumption, if any, would be on the part of the editors: the writer committed the crime of writing. He did not have to be judge and jury. *That was* presumptuous of me. The editors decided the criminality. So persuaded, I let him send the stories out. They were published as stories in magazines, and finally as a book called *The Friendly Persuasion.*

I was visiting at Grace's at the time the book came out, and it was to her home the invitation came to visit New York for two weeks as guest of the publishers.

I considered the invitation ridiculous and dangerous. Ridiculous, now that I had found a publisher, not to stay at home writing more stories. I had started late, as it was.

Dangerous for an invalid to make such a trip. What would happen to my rest periods? My nine o'clock bedtime? My orange juice laced with eggs? My habit of reclining?

"I would faint at an interview, collapse at a cocktail party, die face to face with a live editor," I said.

"You are already dead," Grace told me, "living as you do. Afraid to take a deep breath, or laugh till you cry, or cry till you run out of tears."

"You think I should go?"

"I think you are starting another kind of sickness if you don't. Worse than the other. In the mind, not the body."

There was no hat to kick under the bed this time.

"You send the telegram," I said.

"What shall I say?"

"Whatever you think best."

"You won't back out?"

"No."

"If you don't go now, they may never ask you again."

"I know that."

Grace sent the telegram of acceptance and I departed full of misgivings. A writer should write, not traipse around the country; an ex-T.B.-er should not play fast and loose with the hard-gained remission of his disease.

I had the time of my life, a writer among writers; a

book lover among book people. The pen was not an exotic implement. Neither cocktail parties nor cocktails felled me; editors were human beings. Rest periods were for invalids. I was too busy to take my temperature. I lost a few pounds, which was a good thing.

The trip to my home lay by route of Grace's home, and I stopped overnight there.

She greeted me at the train by saying, "Well, I see that you are still alive."

She made me laugh—as she always had been able, even when laughter undermined my own contentions. My state on returning was unlike what I had prophesied it would be when Grace was urging that I make the trip: that is, prostrate, on a stretcher, en route again to the sanatorium; even I could see that in addition to stamina I didn't know I had, I had possessed the state of mind of an invalid who had been too long institutionalized.

Grace was never one to lecture. Or to say, "I told you so." And she was the last person in the world to think that "mother knows best." She sometimes thought that *Grace* knew best; but she hadn't the least idea that motherhood in itself was a source of wisdom.

She, who had been so determined that I accept the invitation to New York, didn't want the hardihood I had demonstrated there to go to my head. "You're over the hump," she said, "but not out of the woods. Go home and save your strength for writing."

The talk that evening (she came in after I had gone to bed) was of living and dying—though we didn't label it with any such words. I had succeeded in not dying. Would I be able to succeed in something more difficult: living? "Dying is a short horse and soon curried," Grace said. "Living is a horse of another color and bigger."

Grace was sixty-three that evening. Three years earlier a doctor had told her that she had a lump in her breast "which would bear watching." Watching a lump in her breast was the last thing Grace proposed to do for the next few months.

"We'll have a look at it now," Grace told her doctor.

The biopsy showed cancer. In three days, breast *and* cancer were removed—with no one except Eldo knowing of the operation. When I read now of women experiencing trauma because on one side concavity has replaced convexity, I remember Grace. She made herself symmetrical, of course, and usually managed to keep her artificial breast pretty well anchored in its rightful position. Not always, though. Sometimes it rode high, sometimes it drooped dangerously. These gyrations made her laugh. Neither she nor her sexuality resided in her left mammary gland, which had functioned as nature intended it in suckling four children. Her nonchalance no doubt bore as much evidence of Eldo's response to her appeal as to her own good sense. Eldo valued her for more than her bust line—and she had more to offer.

She died seventeen years after that operation of a disease not connected with cancer, but which did affect her memory.

In the week before her death, I, trying to identify myself to her, said, "I'm the oldest of your four children."

"The oldest?" she repeated.

"The one who wrote those Quaker stories."

She misunderstood my words, but her unconscious led her to an insight deeper than my words.

"Oh," said she, "did I get those stories written?"

"Written and published," I said.

"I always wanted to write them. But I married early and wasn't well. It slipped my mind that I did it. I thought I just dreamed I did it."

"It isn't a dream."

"I remember now. The horse race and the son's fighting in the war."

What she thought she remembered was purest fiction, something that never happened. What *had* happened, the clink of her mother's wedding ring as she washed dishes, her grandfather's love of music, the whisper of snow, the rustle of shawls and full skirts in the Meetinghouse on First Day: these, the realities of which she told me, had been *my* dreams. It was a strange exchange. She accepted my fiction as real. Her memories and long-time musings had become my fiction.

"I'm so glad you told me," she said. "I haven't been well, and my memory's got holes in it. It makes me feel better to know I wasn't just a dreamer."

"You weren't," I told her, gospel truth if it's ever been spoken.

There was more I wanted to tell her, but I knew that it was too complex for me to voice.

She had given me birth three times over: picked me a beautiful Indian-nosed father and held onto those bloody strips of sheeting through two nights of July thunderstorm, until I emerged.

The second birth was perhaps even more difficult. One expects a baby to cry, to accept every defeat as final; but a grown woman's determined despair is more exasperating. A baby falls, gets up and tries again. The woman-baby fell twice and decided never to attempt anything as risky as walking again.

Without a life of my own, Grace once again gave me a life, back in "the old country," lost to her in time and space, but vivid in her dream; a land I dared to live in.

"I never could have done it," Grace said, her mind still on the writing and publication she had "forgotten," "except that you encouraged me by listening."

"What else could I do," I asked, "bedfast and you standing over me?"

Grace laughed. "What's the word for that?"

"Captive audience."

"We had some good times in spite of everything, didn't we?"

I kissed her. "Still have," I said.

Part Two

CARMEN

THREE YEARS after Grace's death, I was in London writing the script for an English remake of *Dark Victory,* the picture that starred Bette Davis. My own fear of dying was gone, but not so far gone that I couldn't easily put myself into the shoes of a woman dying of a brain tumor. If the script-writing offer had come from any city but London, I could have resisted it. But London for four months I could not resist. England was not "the old country" in the way Grace's southern Indiana was; but for those whose reading has been done in English and who are bookworms, England has been "the old country" since the first pages were turned.

Lodging had been arranged for me by my employers. It is a fault of mine that I am inclined to like (this isn't always true) what I find myself stuck with. This may be a sign of laziness, or fearfulness; though I wouldn't call Thoreau, born in Concord and determined to live and die there without making an effort to discover anything better, either lazy or fearful.

It was not the hotel itself that so won me, but its

location. It was in Chelsea, on the edge of Belgravia, facing Cadogan Place, and with a side entrance on Sloane Street. The Carlton Tower was near the top of Sloane Street, a street that terminates near the Thames.

A few steps away was a hotel more famous in its day than the Carlton Tower has yet become. Here Oscar Wilde first reveled and was later arrested. Nearby, on Cheyne Row, is the home of the Carlyles; not much reveling there, but no arrests either.

Off King's Road, then the swinging heart of London, on Cheyne Walk, is Chelsea Old Church, where Sir Thomas More worshiped. In 1941, a German bomb destroyed all of the church except More's chapel. The church has been restored and is, for me, the holiest spot in London. I went there the evening after I received Carmen's letter.

The church is furnished with kneelers with embroidered covers celebrating More, and others whose lives—and deaths (a young Canadian firewatcher was killed there on the night of the bombing)—are connected with the church. More's sayings embellish the kneelers. The most famous, I suppose, are his words to the headsman at the time of his beheading: "Help me up, and as for my coming down, I'll look after myself." The most touching, his prayer: "God make me faithful, true, and plain."

These, the words of a canonized Catholic saint, preserved in an Anglican church, have a ring of Quaker sobriety; they make More a man, not only for all seasons, but for all sects.

To the right from the Carlton Tower, ascending Sloane Street away from the Thames and King's

Road, the terrain becomes less religious and also less literary.

Harrods, a few streets away from the Carlton Tower, is the world's greatest department store. It is nearer a world's fair than a department store: more animals are there than in many a zoo; more books than in many a bookstore. And in between these two departments, items that interested me less: groceries and hats and gloves and furniture and musical instruments—but all beyond numbering.

Beyond Harrods is Hyde Park. If Harrods is a world's fair, Hyde Park is a world. Dogs are walked; horses cantered; orators, in what the British call "a sunny interval," reform the world with rhetoric.

Each morning, earlier than I wanted to be awakened, I was roused by the clop of horses on the pavement below my window: the horses of the Household Cavalry were being ridden home to their stables at Knightsbridge Barracks, after early exercise in Hyde Park.

"Sleep," I told myself each morning as that ringing clatter awakened me. But what slugabed would lie snoozing while the horsemen of the Household Cavalry rode by? So, each morning, from my small balcony overhanging Sloane Street, I, whatever the Queen did, welcomed her horsemen home.

The hotel was no literary hangout (though Tennessee Williams was in residence). It had been chosen for me because the producers for whom I worked lived in flats in nearby Belgravia, and they wanted me close at hand for frequent conferences. There was little that was English about the Carlton Tower. The maid, a Yugoslav, was married to an Italian, who was a waiter; they were in England to learn English before opening their own hotel in Italy.

The food was American. An English friend having dinner with me asked, when she saw the size of the steak served her, for the assistance of a waiter. "I am not accustomed to carving," she told him.

At breakfast with Stuart Millar, who had been an assistant producer when I worked on *The Friendly Persuasion* and was a producer here, we were served marmalade in the little plastic containers we had learned to put up with in the States. But Stu had not come to England to eat good English marmalade from small plastic boxes.

"Take these away," Stu, who had not learned to produce for nothing, ordered the waiter, "and bring us Fortnum and Mason's marmalade, either in their stone jar or in a suitable dish."

The waiter murmured something about sanitation.

Stu said, "I am interested in breakfast and marmalade, not sanitation. Bring me marmalade and lots of it, and I'll trust Fortnum and Mason for the sanitation."

We got a cut-glass dish of marmalade and a stone jar, half filled, with which to replace what was eaten.

American tour groups were booked into less expensive hotels; Americans who wanted an English hotel in England went to Brown's or the Connaught. The Carlton Tower clientele was made up for the most part of businessmen, big businessmen from the Continent, talking in deals of thousands of dollars of oil and timber and cars and wines and fabrics.

An African delegation from one of the new republics arrived one day in dazzling costumes. I expressed my admiration to the operator of the lift.

Said he sourly, "Be glad you like them. You Americans are paying for them."

England's empire was crumbling; America, with her Marshall Plan and her advocacy of self-determination, was helping with cash and ideology in the crumbling. We were not loved in England.

It was the day of the shoe with the long, pointed toe. I didn't care about pointed toes, but I did need a shoe that was long and narrow. Long, I could find; narrow, no.

"We have not gone in here for the American craze for long, pointed shoes," the saleslady told me, ignoring the fact that my long, pointed foot, though American, was a fact, not the result of a craze.

Since I arose with the Cavalry, my work, thirty pages some days, was usually finished by midafternoon. Then I began my intoxicating round of the secondhand bookstores in London's West End.

These bookshops were to me what bars are to barflies. I trembled as I neared one. I could smell them afar. I bought a book that listed every bookstore in London, gave its hours, and the nature of its merchandise. I discovered landmarks and "tourist attractions" because they were in the area of a bookshop I was visiting. I found Chelsea Old Church this way; Ellen Terry's and Whistler's homes; the Victoria & Albert Museum. I liked it that way. I like great things to come by chance, not by computer-dating, or map and tourist guidebooks.

In the bookshops were volumes long known, beautifully bound, and at prices much cheaper than those at home: Montaigne and Pepys; Augustus Hare's six-volume account of his life, first with, then without, Mother. There were also books of which I had never heard and which probably had never made their way across the Atlantic; the volumes of Hesketh Pearson and Hugh Kingsmill; of Desmond MacCarthy and Esther Meynell. I discovered also the enthralling anthologies of Daniel George. I am the greatest reader of minor works now living. What distinction is there in having read Shakespeare and Dickens and Tolstoy and Yeats? Who hasn't? They are hurdles one must leap on the way to an A.B.—let alone a doctorate—in English. But Daniel George—who besides me knows him?

On the afternoon of September 3, 1962, I came home to the Carlton Tower laden with, among other curiosities, bargains, and treasures, two more of Daniel George's collections: *Alphabetical Order* and *A Peck of Troubles.*

The overseas mail was in and had been delivered to my room, but before I opened it, I had another look at my books. Everything was going well at home and the reservation for my departure had already been made. I had more to lament for the bookstores I was leaving than for my family at home. I went out to my little two-foot balcony, and, standing there in the late sunshine of the long northern afternoon, I thought, "I am happy." Often you think back to a time when you were happy. Or forward to a time when you expect to be happy. But not often are you able to say, "Now I am happy." I said it then.

Then I went inside, put down George's *A Peck of Troubles,* which I had carried out with me, and picked up the mail. I read Carmen's letter first. I expected instructions about the color of an Irish sweater I was planning to buy for her.

She did not mention the sweater, but spoke of the other members of the family. At the very close of the letter she wrote, "I am going into the hospital on Tuesday, and on Thursday I'll be operated on for a cancer of the lower bowel. There has been no sign of it at all. Dr. Munger discovered it when I had my yearly checkup last week.

"It's very small, and considering Mama's and Papa's experience, I'm not too worried. Though I can't say I enjoy the idea of any operation."

My sister, Carmen, was four years younger than I. She had been put into my arms as I sat in a child's rocking chair on the morning of her birth in Greensburg, Indiana. Had that act given me a feeling more maternal than most children have for their younger sisters? Did I feel more of a mother's pride and responsibility than a sister's?

I began the minute I finished her letter to write to her. It was a difficult letter to write. I felt that I must not express all the sorrow and fear that I felt. Carmen was buoyed up by the good fortune of our parents. Grace, her breast removed at sixty, lived for seventeen more years with no return of cancer.

Eldo, who at seventy was operated on for cancer of the duodenum, with some feet of the upper intestine removed, was still alive at eighty-three; and at the moment briskly wooing a cousin of Grace's who was a preacher's widow. Passionately wooing a brisk widow would be a more accurate description. Eldo had showed me the letters he wrote to Grace after her death. He didn't show me his letters to the widow, but he told me of them and of the poetry he wrote her. And he voiced his fears that he had perhaps been so intemperate in his language that Ivy would expect action as impassioned as his words. He did not want to lead her on.

Edward Dahlberg says of Edwin Arlington Robinson, "He was too moral a man to entice a woman unless he meant business."

Eldo also was moral. He also meant business; but the business of which he was capable at eighty-three might not measure up to what his letters had seemed to promise. He had written Ivy that his position was that of "a man who has discovered a gold mine, but is unable to mine it."

Would this, he asked me, convey to Ivy the information he thought she should have? And not too indelicately? I reassured him, and he sent the letter.

There were people, Carmen wasn't one of them, who thought that falling in love at eighty-three (leaving gold mines, mined or unmined, out of it) was indelicate. I didn't. What can you be doing at that age, unless you're a Thomas Hardy or Pablo Casals, that is better? And Eldo, who wooed at eighty-three, lived on to be ninety; and before that time came, he played many a game of chess with and wrote many an impassioned poem to Ivy, who wisely clung to her widowhood.

With such a father and mother to remember, I could not write to Carmen a cancer-dreading letter. I wrote instead a letter of sympathy for the need of an operation of any kind. For whatever reasons, that is no pleasure.

It is easy to recognize the ways in which Carmen and I differed. We no doubt had likenesses less easy for me to detect.

There were first of all the differences that could be seen. I was big-boned, five feet seven, square-jawed, square-shouldered, freckled, bookish, athletic.

Carmen was five feet five, small-boned, with narrow, sloping shoulders; small-waisted, with the full bosom, taut bottom, and rounded calves that made men who saw her at the beach pretend (she believed) to be photographers searching for suitable subjects for a photographic essay entitled "A Sunday at the Beach." Would she pose?

Carmen believed none of this. Men had a line against which Carmen often warned me. "You believe everything that anyone tells you," she said. Well, I wouldn't have believed anyone who said he wanted me as an illustration for an article on "A Sunday at the Beach." But how do I know? For one thing, no one ever asked me.

Grace never achieved what she wanted: a black-haired, bronze-skinned replica of Eldo. But Carmen, at least, had olive skin and big hazel eyes set in her head with a slant reminiscent of those of our ancestors who had first reached this continent from Asia. She was a towhead, but with no hint of Irish red to contradict the slant of her eyes and the color of her skin.

Carmen and I had a long-standing argument about necklines; neither of us realized until we were in our thirties that each had chosen what was most suitable to her build. Carmen, with her sloping shoulders, oval face, and long, slender neck, preferred and wore high, severe, even mannish collars. She urged me to do the same. She considered my U-necks and V-necks and frills and flounces unchic in the extreme. I, wearing anything like her ascots and Arrow collars, would have felt, and no doubt looked, like a lady horse doctor.

It was less a question of chic than of native understanding of the requirements of differing builds. I needed a softening neckline. Bowler hat and bow tie could not have disguised Carmen's feminity.

We didn't have many years at home together. I graduated from high school at sixteen, and was never at home after that except for visits and vacations. First college, then marriage separated us.

In the years we were at home together, we lived separate lives. My playlife was with my older brother; my worklife was with Grace. Carmen cared as little for hiking and baseball as for housework. She never went on Sunday hikes with my brother and me; never had fencing duels using lathes for sabers; never played burn-out or practiced broad-jumping with orange boxes as incentives for extended effort.

Inside, with two housekeepers already working, she wasn't really needed; and, besides, housework didn't really appeal to her, ever.

So why was it, then, on my visits home, happy as I was to see everyone, that it was Carmen's presence that gilded those occasions?

"Is Carmen here?"

She had for me what is now called "star quality." All of her activities were to me glamorous. They were for the most part going out with boys or getting ready to go out; dressing and arranging her hair; putting on lipstick with a brush and eyeshadow with a practiced finger, and cologne behind her ears.

I looked forward to being with her as I would to being with an actress. But with an actress who was my sister. Not that there was anything theatrical in her bearing or action. Grace with her elocution lessons, I with my debating and literary contests, were perhaps more stagy than she. But Carmen's drama was not a matter of schoolgirl activity, as Grace's and mine had been; her life since the age of ten had been on the real stage, with a real cast of supporting actors: boys who wanted her approval, her company, her hand. The drama was sex; and, except for violence, no theme so absorbing. And the heroine, my sister!

Yet it wasn't for reports of her latest suitors that we sat down together before the fireplace when I came home. She could not have been as enchanted by my life as I was by hers; or by my appearance as I was by hers. But we had common interests: clothes, for one. I couldn't sew and she could; I hadn't the patience to shop for exactly the right fit and the right color. I hadn't her taste, but we shopped together, tried on clothes together, swapped accessories, and laughed at, admired, and were amazed by our images reflected in the full-length mirrors of department-store dressing rooms.

Carmen hadn't my mania for books, but she could understand the nature of such a mania; and she was a reader. At twelve and sixteen we read Elbert Hubbard's *Little Journeys to the Homes of the Great,* and planned to visit each one. She listened with interest when poetry was read aloud to her.

Brought up to believe that dancing was sinful (Carmen, sin or no sin, early discarded this belief), we made up dances of our own which we danced together to the music of the family Victrola: the steps we choreographed were more suited to veld and pampas than to a ballroom, but they were movement to music; movement more exciting even than that demanded by burn-out or broad-jumping over orange boxes.

To this girl (she would always be a girl to me), I wrote my letter of sympathy from London—too bad about the operation—and mailed it.

The minute it was mailed, I regretted it. Your sister has cancer and you write to her as if her trouble is acne or bursitis.

I sent an immediate cablegram, somber and grieving. On second thought, I decided it was worse than the letter. It amounted to a eulogy; fatality accepted. Good-by, dear sister. As it must to all.

I phoned her. Five P.M. in London was 9:00 A.M. in California. Carmen answered. I don't know why voices can convey what pens cannot. Speaking to Carmen, there was neither frivolity nor doomsday in our conversation. The prospect, *that* operation with hospitalization to follow, was nothing anyone would wish for. Nothing for despair, either, remembering Grace and Eldo. Better for a woman, really, if she had the choice, to lose a piece of one's bowel than one's breast. It would show less. And Carmen cared about what showed. She had never been a Grace, a clown, a prankster. A lopsided bust line would depress her; and her husband, too, who was no Eldo.

The conversation ended casually. We might have been talking as in the old days, of necklines and skirt lengths.

"I've written you and cabled you," I told her. "Pay no attention to either. I'm saying now what I wanted to say in them. I'll call Bill the evening of the operation."

I did not feel casual, though. From one's parents one is prepared to receive grim news. Not from one's younger sister. The bell that tolls for them will toll for you, but the bell that tolls for a younger sister is much closer; its knell doesn't stop reverberating, and it says, "You next. You next."

I took a taxi down to Chelsea Old Church; a taxi instead of walking because I was tired after book-hunting; a church, not to pray in, but because I wanted to be where a man whose news had been much worse than Carmen's—and of his own choosing—had worshiped.

"Are you going to a wedding?" the taxi driver asked. I didn't know it then, but Chelsea Old Church was much used for weddings, and had probably been so used that afternoon. The chapel of Saint Thomas More was filled with flowers.

I did pray, using the kneeler, with his own gay words: "'Help me up, and as for my coming down, I'll look after myself.' God help Carmen."

The news after the operation was good. The tumor was small, no involvement of other organs, no need for a colostomy, which Carmen had dreaded. Prognosis: complete recovery with, considering the family history, no recurrence.

Still, there was an uneasiness on my part, to be sitting in London writing of a woman who died of a brain cancer while my sister in California was bedfast recovering from a cancer operation. I was not superstitious. I did not believe that the words I put on paper about Lady Dallas Pember (names for the English version of *Dark Victory* had been changed) would affect Carmen in California. Yet I brought

Dallas to her death with reluctance. I tried to make myself believe that Dallas was a surrogate: she would die instead of Carmen. Nevertheless, I couldn't escape the fact that while I made money in London writing of death and cancer, Carmen in California was meeting the scourge with her own body, no pen to protect her.

I cut my stay in London two weeks short. The script was finished, shooting had begun, and I was anxious to get home.

On my last day, I made a final pilgrimage to Chelsea Old Church. It was raining, but I had been in London long enough now to ignore rain. It was as much a part of the atmosphere as air. I walked to the Thames first, gray river of death and triumph. For how many had it been the road to the Tower? For how many had the sweep of oars carried men to the rewards of fame? More, himself Lord Chancellor, had traveled that way in triumph many a time.

There had been no wedding that day in his chapel. There was no sunshine to set the windows aflame. The memory of the man "faithful, true, and plain" was nevertheless heavy in the place. He chose death rather than abandon his beliefs. He put his head on the block joking: he asked the headsman to spare his beard, since it had done no wrong.

God bless Thomas More. God help Carmen.

Carmen and her husband, while I had been in England, had bought a new house. I was astonished to find Carmen had never liked their old home, which I had always loved. It was small, made of adobe, which tended to melt in the occasional desert rainstorms and to blow away in the more frequent desert windstorms. The date grove (called a garden in the Cochella Valley) shut the mountains that lined the valley's floor on each side from sight. The glory of the valley, which was the morning and evening festival of lights on Mount Tahquitz to the south and on the Pantamint Hills to the north, could not be seen from the house; the windows were small, the trees towering.

Near the house was a combination swimming pool and reservoir, large, deep, bottle-green, reflecting the palm fronds that shadowed it. The pool area was greater than that of the house. This made the house, if you wanted to look at it that way, appear to be a mud hut set amidst a strange tropical forest on the edge of a gloomy body of water.

Carmen had looked at it that way; though the house, in spite of the adobe dust that sifted down and the lack of sunlight (a boon in the desert), had, for me, much charm.

Women can be classified in two ways (a thousand, I suppose, actually): those who, on the arrival of unexpected guests, look in a mirror and refurbish themselves; those who grab a dust mop and refurbish their homes. Carmen looked in a mirror. Like a Mary, as contrasted with a Martha, she chose the better part. She talked with an entranced guest while Martha sweated away over the mop pail and dustpan.

But the dark house with Carmen's arrangement of its furnishings was, I thought, charming. In this, as in her own choice of dresses, Carmen could not be dull or stodgy.

Nevertheless, Carmen had been glad to leave the adobe house under the date palms for a new house two or three miles distant, on the first slopes of the Mt. Tahquitz foothills. Here all was visible: the mountains, the sky, the desert floor, the great, manly (womanly, actually) date trees. Nothing is sexier than a date garden. All of its trees are 100-percent female, all artificially inseminated by the implantation amidst their female fronds of segments from male trees.

To her new house, after I had put my own home in order, I went to visit Carmen. Her operation was only three months past. She was somewhat thin and pale; but the cause of the operation she had put out of her mind, or at least out of her conversation. She had Grace's and Eldo's example and her doctor's assurance to make her optimistic.

Eldo came then to visit her, also. He said to me wonderingly, "It doesn't seem like this should have happened to Carmen." It didn't; though I didn't know to whom it *should* have happened. Tragedies are for heroes with tragic flaws, not slender, pretty younger sisters. Flowers must die; but they don't deserve cannonballs.

Eldo said to me, "Your mother saved *your* life by prayer."

I didn't know that; Grace had never told me. Who, if the need arose, would save Carmen's life by prayer? I think Eldo believed, and rightly, that neither he nor I was up to that job. And Grace was gone.

"I liked your old house better," Eldo told Carmen. "It was a real shelter. Don't you feel like a fish in a bowl here?"

"Would you rather be fish in a bowl with glass sides, or a fish in a windowless bowl?"

Eldo thought about that.

"The fish looks out?"

"Why do you think they have such big eyes?"

When Eldo left for a nap, Carmen said, "After the first few years, we could always have come to a place like this. But we had planted those trees and built that house and dug that reservoir, and to leave it seemed heartless. Especially to Bill. But he was almost never there in the daytime. When he came in at night, it was too dark already to see out. Lights were on, fire burning." (Desert nights can be icy.) "It looked cozy to him. I was there all day. It looked gloomy to me. Depressing. A tomb."

She didn't like that word. "O.K. It did. Compared to it, this is the resurrection. All brightness and light and lights changing colors. And the world let in, not shut out."

"This doesn't stay as cool as the adobe house, does it?"

"Haven't they heard about air conditioning up north?" Carmen asked. "You don't have to build to keep the sun out any more. You can let the world in and turn on the air conditioning."

There had been an unseasonable early fall rain, and the gray desert and foothills, except where the date gardens covered them, had the faintest tinge of green; like a rat, dead but unburied. It was the end of an autumn afternoon. Eldo was sleeping. Carmen and I, with a gimlet and Scotch-and-water, seated ourselves where we could see the sun go down behind the Pantamint Hills. Drinking offended Eldo. Grace had long since decided that none of us were cut out to be drunkards, and, short of that, what did alcohol matter? Alcohol was both a sedative and a laxative, and hence medicinally useful. Not so Eldo, now under the influence of Ivy, the preacher's widow, so Carmen and I kept bottles from his sight.

"Girls," said he to us who would never be girls again except to him, "have you ever thought of joining the A.A.?" We, with our drink before dinner, and glass of wine with dinner, would be an embarrassment to the A.A. We let him know that we knew of the organization and would turn to it at the first sign of addiction. This comforted him.

The sunset boiled up all colors: from fire's red blaze to the gold and apricot of ripe peaches as it faded. The whole stretch of the great valley was visible to us. Earth flesh in the desert is not Victorian. The desert is not prudish. She is bare to the eye, naked as her inhabitants when undisguised by their skirts and breeches.

Carmen said, "I don't eat my heart out about it. Or I try not to. I remember Papa's and Mama's luck. But I remember, too, both grandmothers dying screaming."

"They didn't have your early operation."

"I know that. But I don't intend to die screaming."

She wouldn't. Both Grace and I "took on" when in pain. We were Irish. We didn't wait for the wake to wail. We wailed while we were still hurting, not leaving all the work for others after we were past helping. Carmen, with her olive skin and slant eyes, was as silent in pain as her forefathers from whom she had inherited them.

She would not die screaming.

Carmen and I were together that Christmas and again in the spring.

In the spring, Eldo, who had been unable to persuade the widow to marry him, had been persuaded by her to move to the Quaker retirement home in Oregon where she lived. This seemed the only legal way in which he could live under the same roof with her; and to live any other way was unthinkable to both of them.

So Carmen and I went down to help Eldo break up housekeeping before his move.

There we walked into the past; a past more distant than our mother Grace's; a past of furniture belonging to her great-grandfather; of pre-Civil War marriage licenses and of pre-Revolutionary letters from the Friends Meeting in Dublin telling Friends in Pennsylvania that these Irish Quakers came with clean bills of health: finances and marriages all in order.

Quakers had learned to hang onto such things through necessity. Since their marriages were illegal ("I, John, take thee, Mary"), their births and deaths had not been recorded in England or Ireland. They kept their own records and carried their papers with them as they moved westward.

There was, in this house where Carmen had been married, I had lain ill, and the screaming grandmother had died, much that was pure Grace. There were the dish towels she had made during Depression days from the cloth sacks that had originally held rabbit food. There were Grace's scrapbooks, a stack of them, waist high and half the size of a card-table top. They contained, surprisingly (when we had come home we had always been too busy talking to look at them), more items about the history of the world than the family history. If all other records disappeared, the course of World War II could be pretty well followed by reading those scrapbooks.

There were dishes, whose history we knew: wedding and anniversary presents; a glass plate with a Liberty Bell in bas-relief celebrating the first hundred years of the Republic; cake stands and covered compote dishes large enough to hold enough stewed pie plant for a family of eight. Autograph albums. "When you and Eldo are wed remember that nothing don't chew like bread." A rejected suitor wrote that and Eldo never forgave him.

Coal-oil lamps, crocheted tablecloths, pictures pasted on the inside of cupboard doors, the Hohner harmonica. Grace was everywhere.

In the kitchen was cooking equipment of all kinds, for Grace had been a great though slapdash cooker. There were homemade steamers that had originally been Hills Brothers coffee cans in which Christmas plum puddings were cooked. There were containers in which duck eggs had been put down in waterglass to preserve them. Mason jars, filled and unfilled. Notebooks full of recipes that once recorded were ignored. Grace cooked "by guess and by gosh," but her guesses were good.

There was her huge aluminum tamale piepan. Who would ever use that again?

When she made tamale pie, Grace quoted a risqué rhyme of her girlhood.

> *"Tom and Mollie sat on the sand*
> *Telling jokes and jollies.*
> *The sand was hot to Tommie's pants*
> *And also hot to Mollie's."*

This rhyme was naughty because it mentioned an unmentionable: a girl's underpants. Southern Indiana, though it didn't mention ladies' undergarments, knew that they existed. I was surprised that tamales were known to exist there at that time.

Grace was pious and prayerful, but she liked jokes and jollies.

I went to the upstairs bedroom where I had spent my long tubercular months. The doves still called from the palm trees that lined the driveway. The bed under which I had kicked my hat was still there. I lay down on it and looked out to the oil tanks on the distant Coyote Hills. I could remember invalidism; but I was separated from those years, not only by time, but by a body that functioned in an entirely different way: a body that took life for granted.

Downstairs, Carmen stood in front of the fireplace and tried, I suppose, to remember herself standing there as a bride.

What did Eldo remember? Or try to forget? He had written to Grace in the first months after her death letters that he sent to me, in which he said that he would live on for the rest of his life, as they had planned, in this house where they had been so happy.

Grace had known better. She gave, before her death, many of her belongings to Carmen and to me. "There'll be another woman in here after I'm gone. She won't know the history of things. They're all Milhous and McManaman, and Eldo doesn't know their history, either. You girls take them now. They're yours by blood."

Grace wasn't bitter about the "other woman." She knew that a man as devoted to her as Eldo had been needed a woman-centered life. She would have laughed at some of his impassioned poems; but she would have recognized that love and hope are better than love and loneliness.

Carmen and I felt like Chekhov characters in *The Cherry Orchard.* The land that our Hoosier grandfather had bought, sight unseen, as an investment had been a barley field. Then orange and lemon trees had been planted, had borne fruit that, except for the Depression years when the fruit was burned, had sent children to college, paid for weddings and hospitalizations.

Now the trees, still alive and bearing, were being bulldozed out. Eldo was subdividing the ranch. First sagebrush, then barley, then the gold and yellow of oranges and lemons; no more Hoosiers and Buckeyes and Sooners and Okies. The westward tilt that had put *us* there in the beginning was bringing more Easterners. Could we blame them?

Eldo, writing letters to Ivy, saw it all, I believe, as a means which would happily place him by her side. It was a beginning for him. An ending for us. We could no longer say, "We're going home," or walk through the orchard hunting for the two well-hidden tangerine trees.

Out in the big red barn that had once sheltered horses and cows and cultivators and disk plows and Hupmobiles and tractors was stored the Victrola that had provided the music for Carmen's and my early dancing. With the advent of radio and television, it was no longer used. But it was too big, too ornate, and reminiscent of days past to be sent to the junk yard.

Carmen and I put a record on, "Aloha, Farewell to Thee." The machine played as well as ever.

"Let's dance."

Carmen had learned to really dance: fox trot and waltz and tango and rhumba. Maybe the twist and the jitterbug, too. It is impossible to discard technique once learned. The primitive can learn; but the learned, be it in dancing or painting or writing, are false when they assume a primitivism they no longer possess. Grandma Moses was doing the best she could. There is truth in what she put on canvas. A Wyeth cannot imitate her unless caricature is what he intends.

So Carmen could no longer plunge and leap as if grass-skirted or G-stringed. I had never learned the steps of a structured dance. I could do nothing but leap and plunge. We were no longer ideal partners, but we danced together. Outside, the trees, alive, the sap still running in them, Valencia oranges, already yellowing, were being uprooted. Inside, the old man wrote love letters; in the unused barn the two middle-aged women danced. "Aloha, Farewell to Thee."

The Chinese, ignoring perspective, of which they are not ignorant, paint from what they call "the angle of totality." Since we know that there are trees on the distant mountain (even though from where we stand we can't see them) and know that (invisible to us) a mockingbird sits there and sings, why not put them in the picture? What is an artist for if not to reveal the unseen to the onlooker?

I could hear the bulldozer, the Victrola, the shuffle and clop of Carmen's and my feet on the barn floor. But I wanted also the sound of Eldo's unseen pen boldly filing gold-mine claims.

Most of all I wanted a sound that rang only in my

memory's ear: Grace slapping pies in the oven and doing imitation bird twits in response to her pet canary, the only animal she ever really tolerated.

There was no angle that would give me *that* totality, visible and invisible, past and present, audible and inaudible.

"Do you really think Mama's prayers saved you?" Carmen asked.

"I don't know. I expect she prayed."

"Do you really think prayer can save the dying?"

"Some people. Some prayers."

"You?"

"No."

"Me either."

I did not see Carmen again until the Fourth of July. On that day, she drove Eldo, who had left the Quaker Retirement Home in Oregon for a week to visit her in Southern California, to my home in the north of the state.

She was driving a new Karman Ghia.

"Did you sell your Thunderbird?" It had been her pride, a yellow convertible.

"No. But I couldn't resist a car that sounds as if it were named for me. Besides, you can't take it with you."

We excuse extravagances that way all the time. Was Carmen saying something else?

I helped Carmen with her bags into her room. "It's so dark."

"This has always been your room."

"I forgot the vines. I can't see out."

The room's casement windows face west, and though there is never desert heat here, 105 degrees is

105 degrees, even when only fifteen miles from San Francisco Bay.

Wisteria vines hung in a thick canopy from the roof, shading the room from the afternoon sun; and, yes, shutting from sight everything but the driveway that led to the house and the dense growth of oak, madrone, and manzanita on the far side of the drive. No sun, but also no sky, no clouds, no stars, no valley of vineyards or line of hills enclosing the valley. Green, green. A room like an undersea cave.

"The vines should have been trimmed," I said.

"It's nothing. I've got a mania for seeing out recently. Did I used to have it?"

I couldn't remember.

She had brought a closetful of clothes: dresses for wear in the dressy town of San Francisco, for dining out, for the theater.

"I thought we might be going into the city," she said.

"We ought. I'm a stick-in-the-mud. Make me go. Do you want to rest a little before dinner? You've driven about six hundred miles today."

"I'll take a nap. I didn't sleep much last night."

"I didn't remember about you and wanting to see out—but I remember about myself. I hated the closed-in greenness when we first came. You and I were brought up in the stubble of barley fields to begin with. Near at hand were foothills; then beyond them the blue mountains. You could see clear to the edge of the world. We were happy children there. I think that if you were happy when you were a child, you want the same landscape again. We'll miss openness all of our lives."

"Do you think that's it?"

"It's the way I've figured it for myself."

A cat jumped up on the window sill, a descendant of Samantha's named Private Eye.

"How many cats do you have now?"

"Too many."

"I tried to change that at home," Carmen said. "You remember Nippy? She'd had about fourteen batches of kittens. The last four were darling—but I decided Nippy should go to sleep. I gave her enough sleeping pills to kill a tiger."

"So Nippy is no more."

"Nippy is fine. The kittens are dead."

"How did that happen?"

"Nippy threw up the pills. The kittens ate them."

"Poor kittens."

"They taught me something. Don't let me nap more than an hour. I won't sleep tonight."

I left her in the gloomy room and went to take Eldo his favorite tipple: a frosted root beer.

I should have given Carmen Eldo's room. It's smaller, has plainer furniture and a drabber rug, but its windows open out to all there *is* to open out on: lawn, driveway, and trees. You're shut in there, too, but by the nature of the world itself, not by vine-covered windows.

Eldo had been tired by the six-hundred-mile drive if Carmen hadn't. He had been lying down but sat up to drink, and when he had finished, he said, "Revive us again. Where's Carmen?"

"Taking a nap."

"She isn't well."

"How do you mean?"

It was a question I could hardly ask and he could hardly answer. Then he said, "Yes."

"The operation?"
"Yes."
"Did she tell you?"
"No."
"What makes you think . . . ?"
"You know how hot it is at her place now? She sleeps with an electric pad wrapped round her stomach."
"Because of pain?"
"Well, it's not to keep warm."
At that moment there began ten days of sorrow, more bitter in their uncertainty and determination to hope than a verdict of doom and hopelessness.
I tiptoed down the hall to Carmen's room. She was sleeping with an electric pad held close to her body.

After I had been with Carmen, under the sentence of death of her new symptoms, for a week, Carmen said that the doctor told her that her pain and her inability to have a bowel movement were the result of scar tissue at the rectum. She believed it to be a return of the cancer. It might be so. And, in any case, I well knew that believing it to be so is as bad as the fact.
The sudden assumption of the coming fatality of another is as unexpected as falling in love. I could not separate myself from Carmen: suffering, examining the significance of her symptoms, facing death and facing it in one of its most horrible forms.
The four years' difference in age, a quasi-motherly obligation reinforced, I think, many times by Grace, who spoke to me of my duty to Carmen; not only the age difference, but something in my character and Carmen's made Grace think that I should be an ex-

ample to Carmen. This riled Carmen, and has, I expect, placed upon me an unnatural feeling of responsibility.

Added to all this was my innate response to Carmen; to her particular attractiveness. She had a face and manner I liked. I thought she was beautiful. It was one of my greatest pleasures to make her happy: to loan her my clothes, my money when I had any; to have her bring her boy friends to our house and to prepare for her the food she liked.

Yet the time came when I was older and less well, when I refused to function any longer as a quasi mother. I had taught her that she could depend upon me and demand of me. Then, suddenly I asked her to be a sister, not a daughter. She couldn't do it. I shouldn't have expected it of her and should not have let the fact that she couldn't make this response cause me to condemn her. My whole happiness with Carmen lay in unconditional loving and giving.

If there is any lesson for me to learn from my life so far, it is that love must not abate. It must not hinge upon reciprocity. If you have truly loved, love on, no matter what. Admit that this act or word in the loved one is undesired—but love on. Otherwise you build up for yourself great suffering.

For some reason I ceased to live in the center of Carmen's fear and sickness. It moved away from me, or I from it, for reasons I didn't understand. Her sorrow and illness moved out where the world's other tragedies and inevitabilities were: with Negroes beaten and Indians starving and old people abandoned. How it happened, I didn't—and don't—

know. Perhaps you can hold your heart only so long against the sharp point of another's suffering. Something requires that initial suffering: sense of guilt? Masochism? I don't know what. But gone finally.

When death threatens, when a good-by is faced, how one searches the past for images, begins to shoal up the past for future use. I remembered Carmen in Hemet, listening on our first good radio to "Ah, Sweet Mystery of Life," and saying, "That is my favorite song in the world. Oh, dark mystery of life."

I used to have a pain in the throat, a constriction in the chest which I called sadness, without knowing any specific cause for it. Now I had the same pain, the same constriction, and knew the cause: Carmen's sickness.

When she was hopeful, when her symptoms let up, when we laughed and talked about old times, the ache in my chest disappeared.

I went into Eldo's room at noon. Carmen was there on his couch. She began to cry when I spoke to her. She had had pain all night, had taken two painkillers since breakfast that hadn't helped. It hurt so to see her, hear her; Carmen, the pretty one, the gay one, the little sister.

I went to town and bought champagne, malfattis, chocolate pie, mile-high with whipped cream, rum cookies, *Vogue, Harper's Bazaar,* movie magazines: stuff calculated to be abrasive to neither mind nor stomach.

When I got back, Carmen was better. She had been swimming, the first time since her operation.

We ate and drank. Eldo listened to the Giants-Philadelphia game and ignored our bottle.

To be with Carmen, follow her every step of the way in what she had to suffer—that was my ideal of the way I should go. I couldn't do it. I didn't even know that she wanted me to do it. A duplicate in suffering is of no real help to the sufferer. I was able for so long to believe that all was well with Carmen, enclosed by her tamarisk hedge, under the desert stars, inside her adobe walls.

Those good times were perhaps gone forever: the confident years when the family was intact and we were all young, Grace and Eldo, too, together.

For the remaining days of Carmen's visit, she was better. No pain, hopeful once again. "The champagne and the *Vogue* did it," she said. "That's the medicine I need."

We never got to the city. The party dresses hung in her closet unused. You can gauge a woman's character, or at least the strength of her imagination, by the clothes she packs. Much that she will never use in that folding bag? She has not given up: the present is drab, but the future is full of promise.

Max's intentions when Carmen left were good, of the very kindest. He wanted to make her send-off gay, gala. The first dahlias were blooming. He decorated the Karman Ghia with them. On top of the instrument panel he put yellow roses

If the flower-decorated car was an omen to Carmen, she didn't show it.

"Harry Maxwell," she said. "My favorite brother-in-law."

She kissed us both good-by. She went out of the driveway fast, in her usual gravel-spurning way.

Eldo, who was standing with us, said once again, "It doesn't seem like it should happen to Carmen."

I was in the East until the end of August. Before I got home, I received a letter from Carmen.

It began, "Sister, dear sister, come home and help me die."

She had been to the hospital and had been told that scar tissue was not the cause of her pain. There had been a recurrence, inoperable. Cobalt had not helped.

"Do you know what a nurse said to me? 'Do you think you should be smoking so much?' Think of that. What was she afraid of? That I might get the habit? Cancer of the lung?"

I phoned her at once from Oxford, Ohio. I was crying.

Carmen was as commanding about that as she was about other matters. She hadn't been born in August for nothing, she a lion and I, July-born, a crab who travels sidewise.

"If you're going to cry, I don't want you around. Bill and I've shed all the tears we have to shed. Crying is past. What is, is. I know what I'm going to do and you can help me. Bill can't do it. You're the only one I've got to help me. But if you're going to go around moping and crying, I'll get along without you."

"I won't cry. I promise you. I won't."

I stopped at that minute and Carmen never saw me cry but once again. Actually there were not many more tears shed. The July agony of alternate hope and despair had passed.

All of us know that someday we will die. We do not cry about it. If there were a chance that we would not die, then we might be on the rack; might not be able to stop hoping and wondering and fearing. Carmen had been taken off the rack. There was no more hope for life. There was now an energetic and almost gay period of planning the kind of death she chose: the kind of death and at the time she chose. *She* had not chosen death: fate or genes, or God, which is perhaps the name we give that combination, had made the choice. She at least intended to have some voice in the matter. Why should she spend two or three months in agony? Or so drugged to avoid the agony that she no longer had any existence as a human being? What God would want His children to die in that way? What child of God would so malign his Creator's nature as to believe that such a death was His choice?

I was met at the airport in Palm Springs by a sister-in-law of Carmen's. The temperature was 115. Part of the wonder of Palm Springs is its greenness in the midst of such heat; its unnaturalness, like finding moss growing in an oven. The wonder is accomplished by water. A desert is not made by heat but by a lack of water. Put water on sand and you'll have a jungle.

I told Hazel, as we drove through the town, about Carmen's encounter on just such a drive with a visiting Belgian youth. André was twenty-one, one of a contingent of European young people taking part in what was known and is still known, I think, as the Experiment in International Living. I don't know how the experiment has fared elsewhere; in Napa, that year, the entire group of three was finally housed with us, internationalism proving unendurable for more than a week with the other hosts.

This was easy to understand. André, our original guest, had been a boy in Belgium during the German occupation and had been taught by them that all Americans were crude, uneducated, and vulgar.

Max and I took him to visit a friend's newly built but still unfurnished home.

"This," said the host, of one room, "is the future library."

André hastily retired to another room from where we, or at least I, heard barely suppressed snickers. I went to him at once.

"What is the matter, André?"

"An American," he said, choking with the idea of the ridiculousness of it all, "with a library."

To me he once said politely when talking with others and having mentioned the Crusades, "You have perhaps heard of the Crusades?" He thought it unlikely, but I had fooled him a time or two before, and he was being more careful.

He was rich as Croesus. His family owned plantations in Africa as we might own a lot in downtown Napa. But he had never owned or driven a car, and he could not keep his hands off mine. He knew about drivers' licenses and stayed off the highway; but in and out of the garage and up and down our driveway he went, taking part of the garage with him and leaving part of the car behind.

When I spoke to him about this, he reminded me that G.I.'s while in Belgium had wiped their feet on priceless tapestries, works of art that could not be replaced; while neither car nor garage were works of art and both could be replaced.

"It takes money, though."

"You and your husband both work," André told me, "and have the money for the repairs."

That *his* money should be used for the damage he caused was unthinkable. *His* money was old money, family money, money with a history, money that he had a duty to hang on to. Ours came in pay checks for teaching and writing. It was without history and meant to be spent.

We had taken André to Palm Springs to show him the wonders of the desert, and Carmen had taken André and me to lunch in that town.

After lunch, André said, "Many movie stars have homes here, don't they?"

"Yes," said Carmen, "they do."

"Would it be possible to drive past some of them?"

"Of course," said Carmen.

"In Europe we know that movie stars are the American royalty and that Americans worship them. I would not like to go home without seeing where they live."

"Certainly," said Carmen.

Now, I knew that movie stars had homes in Palm Springs, but who the movie stars were or where they lived, I had no idea.

Carmen seemed to be informed. She whipped her Thunderbird up hills and around hairpin curves, pausing before mansions shaded by palm and olive and oleander, which were reflected in pools of blue water.

She was as knowledgeable as a tour director. "This is Norma Shearer's home. . . . Here is where Alan Ladd lives. . . . Bing Crosby was one of the first to come here. . . . Esther Williams, the swimming star, has that especially high diving platform. . . . This is Humphrey Bogart's hideaway."

"It is as I expected," André said. "The movie stars live like kings and the Americans worship them."

"There are more," Carmen told him. "But now I must go home."

When she was out of town, headed toward their own date garden, Carmen stopped the car.

"André," she said, "not a word I told you was true. I don't know who lives in those houses, or whether any of those movie stars live in Palm Springs."

"Why did you do it?" said André, pitifully capable when the occasion demanded it of being the poor boy far from home instead of the superior European.

"To teach you a lesson. You are a big boy now and ought to start thinking for yourself. O.K. You believed what the Germans told you—but you were a kid then and didn't know the difference between fact and propaganda. Nobody is propagandizing you here. You can judge for yourself. If you haven't enough sense to see that my sister and I aren't movie-star crazy, this trip is wasted on you. You better go back home tomorrow."

Hazel and I laughed about that. We were on our way to where that outspoken woman was dying, but we remembered with laughter what she had said and done in the past.

"Once I was taking care of Carmen after her operation for appendicitis. She asked for a soft-boiled egg. I hard-boiled it by mistake and she threw it at me," I told Hazel.

"We could do that," Hazel said. "Or talk that way to André, and it wouldn't be funny. We *look* as if we could do it. And Carmen doesn't. It's like seeing a hummingbird attack an eagle."

Two months passed since I had seen Carmen. She was in bed, but she looked better than she had in July. She had on lounging pajamas, make-up; her French knot was as elegant as ever. The big sunlit room was alive with the fish-shaped dancing shadows of olive-tree leaves. In the patio outside her room, desert quail were feeding and calling.

"I thought I'd never see you again," Carmen said.

"You knew I would come."

"I didn't know whether I'd be here or not."

I couldn't ask her about that; and I didn't have to. I knew what she meant.

"You look better."

"So far, there's a pill for every pain. A pill to make my bowels move, and one to stop the pain when they do. A pill to ease the pressure of the damned thing on my bladder and a pill to put me to sleep; then there's a pill to relieve my depression when I wake up. There's a needle with God knows what in it to use when none of these things work, and you'll have to learn now to use it. I've got a nurse-housekeeper for daytime, and I thought you could be my night nurse. The night nurse doesn't have to do anything but use the needle when things get too rough. And when they really get too bad for *that*, I won't need you any more."

I had a room separated from Carmen's only by a hall and a bathroom. An electric bell had been installed so that she could ring for me when she needed me. She wasn't bedfast, could walk to the bathroom or out onto her enclosed patio, but these trips pained her, and she didn't leave her bed very often.

I unpacked in my room with its view across the valley floor to the bone-gray hills on the east. It was nearing sunset and the gray was turning rose-red.

This was Grace and Eldo's wedding anniversary. What would Grace have thought sixty-two years ago if she could see her daughter today—dying? Refused to conceive her? What would Carmen choose? Life, with this death? Or no life?

It's as Eric Hoffer says. All change is an ordeal. Hoffer suffered as a young workingman when he had to shift from picking peas to picking beans. He was filled with dread. How much greater the ordeal when the change is from life to death, from the known to the unknown. Though the ordeal of change is made easier when the known is unendurable pain.

It was a hot, clear, lemon-colored twilight, a hot, dry wind whistled around the house. Beautiful stormy sound. Carmen said she loved it too. I hadn't known that.

It is easier to be with, and to try to help, a sister who is suffering—and dying—than to be away from her. The days together provide a series of small crises, or even small unexpected pleasures, instead of the one big abstract tragedy I had experienced, away from her. There is great rest for the heart in trying to make someone easier.

Carmen said, "Pain is a blessing, because it makes death a separation from what is so bad."

I wasn't sure if Carmen would want me around constantly. If I were dying . . . when I am dying . . . do I want relatives near? Someone who wants to be wanted? That's a great burden. I didn't want Carmen to feel that with me. I needed Grace, but I was no Grace and Carmen's disease couldn't be cured by faked temperatures and augmented orange juice.

"The suicides are then a group apart, technicians, planners, plotters." So writes Elizabeth Hardwick in *Seduction and Betrayal.*

Carmen and I became planners, plotters, technicians. We knew what had to be done. We were *pretty* sure how to do it. We intended to be sure. There were those who might try to prevent it. Those we must circumvent. The timing must be right. We would have to decide about that. Death was the goal, but Carmen's kind of death, not nature's savage torture system. Not a slice at a time, the feet in the fire and the testicles squeezed off by the tightening clasp of a drying deerskin pouch; More's death instead, self-elected and not prolonged, was the model.

We were not as joyous as technicians and planners building a ship for launching; or as writers composing a novel whose incidents will reveal (as a life lived cannot) the human condition. We were not prospective mothers carrying babies and eager for their birth. Death was what we plotted. We were not eager for it; but we were eager to make the ordeal of change, which had been decreed, as easy as possible.

Overheard, no one would have guessed from the tone of our voices the nature of our subject: suicide. Death by one's own hand—the crime for which the suicide not too long ago had been denied Christian rites at burial; buried instead at the crossroads with a stake through his heart. An act still illegal for the person who gives assistance to the suicide. Jumping off the Golden Gate Bridge is not, for the jumper, punishable, even if he survives; but if the faint-hearted jumper brings along a friend to give him a shove in case his courage fails, the shover will be in immediate trouble.

Carmen had at first declared that the act upon which she was determined must be a solitary one. She must do it alone. Later, when she wrote me her "Sister, dear sister" letter, she had changed her mind. "I want someone to hold my hand at the last," she said. "Bill can't do it. There's no one but you."

She had originally planned the ordeal of change for the night of the first Sunday after I arrived. Her son would have been there for the day. I would be present. Why postpone? But the day had gone well, laughter, little pain. She faced the evening with dread.

"You can plan the means with your head," I told her. "But you can't plan the time with your head. You will have to wait for your body to tell you the time. Your hand will know when the time comes. It will reach out for what you need and you can't stop it. Is it reaching tonight?"

"No. It is dreading."

"This is not the night."

There were two beds in Carmen's room. I slept down the hall from her with only a bathroom separating us. Bill, since his sickness and Carmen's, slept in the bedroom off his office in another wing of the house.

My room was connected with Carmen's by a loud buzzer that she could sound if she needed me. She had a shot at eleven o'clock when I went to bed, another at two, and a final shot at five. Carmen believed that it was better, and that it took fewer shots, if she had the painkiller before the pain started rather than after. For this reason, I set my alarm clock so that I would awaken without Carmen's having to buzz for me. Oftentimes I would have to awaken her to give her a shot. This, of course, did awaken her, and frequently she was hungry. I would bring her a custard baked by the day nurse for this purpose, or make her a cup of Ovaltine. She ate and smoked and we talked until the medicine relaxed her enough to let her sleep.

"I like to wake up in the night and smoke and talk. Lots of married couples do this"—this was news to me—"but I never could with Bill. If he was awakened in the night, he couldn't go back to sleep again."

I could sympathize with Bill. I had begun to take a sleeping pill myself after my nightly sessions with Carmen in order to sleep again—although I didn't tell her this. She did tell me, "Don't go to the bathroom every time you wake up in the night." I had been doing so out of some conviction that I was thus killing two birds with one stone. "You'll train your bladder to be emptied every two hours and you'll have to do that the rest of your life."

I decided she was right, and got, though it took a little effort, my bladder retrained after a few nights.

We talked a lot about Grace. We saw her from different perspectives, neither from the angle of totality. Carmen was her father's girl; I was my mother's helper. Eldo's bent with me was professional. He would immediately and without urging plunge with me into the intricacies of nationalizing the railroads, if that was the subject I was debating on; or try to unravel Browning's *The Ring and the Book,* if that was the poem I was studying. But I was unable, as Carmen was, to sit on the arm of his chair and brush his hair. I was, as Grace's helper, "the other woman," not "Papa's little girl," with Eldo.

Grace, with boys constantly pounding on the door asking for Carmen, was more concerned about Carmen's "getting into trouble" than about my doing so. She lectured Carmen more than she did me on the subject, and Carmen resented it. An unmarried daughter would have been as sorrowful to Grace as a Mongoloid child or a dwarf. She knew the joys of marriage, and if anything she could do would help find us husbands, she would do it. But no playing around by us before the knot was tied.

I was engaged at seventeen to a responsible, curly-haired Quaker; but Carmen at twenty-four was still flitting from boy to boy, dating two boys in one evening, and half a dozen in a week. She went to public dances; she saw risqué plays like *White Cargo.* She was young enough to be a part of the Roaring Twenties. I was young enough, still a teen-ager in the twenties, but my effective sex-training had been as pre-World War I and pre-Californian as Carmen's hadn't been. I had the inhibitions of small-town Middle West.

Drinking her Ovaltine after a shot one night, Carmen said, "You always wanted to please Mama, didn't you?"

There was no use denying it. I had.

"I wonder what she would think of what you're doing now?"

"Now? Approve. One hundred percent."

"She found fault with all those Wests and Clarks who hanged themselves and cut their throats and drowned themselves at the drop of a hat."

"That was it. At the drop of a hat. Don't die today for what will seem tomorrow no more than a molehill."

"I ain't got no molehill," Carmen said.

"I know that. So would Mama."

"She let her own mother die screaming."

"Can you remember that, Carmen?"

"I can remember."

"You were only five."

"I remember."

"She was sorry."

"How do you know?"

"She told me so. She said that if she had it to do again, she would save Grandma those last two months of anguish."

"How?"

"She didn't know how. That was the trouble."

Carmen knew how. She had accumulated during the last six months, and not from her own doctor, a hundred capsules. She kept them in a large jar, intended perhaps for candy, square, with an ornate lid that, ground glass to ground glass, fitted the jar tightly. The roseate capsules seemed almost to shed light; to gleam like a votive candle. The jar was kept out of sight deep in a drawer behind lingerie. But Carmen cherished it like a miser his coins. She liked to look at it occasionally, to make sure it was there and the contents untouched. It was a reverse crucifix. It promised her not life everlasting, but death when she asked for it.

When I began to need some sleeping pills to get me through the night, Dr. Alfred Munger, Carmen's doctor, a man who looked like Abraham Lincoln, wart, beard, and all, prescribed some three-quarter-grain Nembutals. They were yellow and without any of the fox-fire gleam of Carmen's big, strong, bloody death-dealers. This pleased Carmen. It gave her confidence. One thing for my naps. Another for the sleep everlasting.

This was the word used instead of death: sleep. Carmen was going to sleep. She was not going to kill herself. Cancer was going to kill her. She planned to be asleep before that happened. But though she used the word "sleep," death, she knew, was what she planned.

"What," I asked, "would you do if all the pills were stolen?"

"That's not very likely."

"But if they were? You couldn't collect another hundred in time."

"Not unless I went down and held up the local Rexall."

"I don't see you doing that."

"Me either . . . Would you shoot me? Bill's got plenty of guns. It would be an act of mercy."

"I'd probably just put out an eye or something."

"That's what I'm afraid of. You could buy rat poison. They'd trace that to you, though, and put you in jail."

"Lots of good books have been written in jail."

"I don't know that rat poison'd be much better than the damned thing I've got. Rats with poison would plead for some pills. I've seen them die."

Remembrance of my san plan came to me then. "What I was going to do at La Viña when they told me I was going to die was to get into a bathtub of water, then lift a turned-on electric heater into the water with me."

"Powie," said Carmen. "Electrocution. Like Barbara Graham. Could you do that?"

"Carry you in, do you mean? Put the heater in? If you were screaming with pain, I think so."

"If this gets much worse, I'll be screaming."

"I could do it, then."

But the pills wouldn't be stolen, and the real problems had to do with getting them down secretly, and of keeping everyone away until they had taken effect.

After about two weeks, I moved out of my own room and into the twin bed in Carmen's room.

One night I had been awakened by the pressure of a hand on my arm. My alarm clock was rattling itself to pieces. Carmen was sitting on the edge of my bed.

"Why didn't you ring your buzzer?"

"I did. I have for ten minutes."

"I slept through that and this clock too?"

"You're not awake yet?"

"Is anything wrong?"

"I'm hurting. I wanted a shot."

I sat up in bed and began to cry, the first and last time while I was there. Then I remembered what Carmen had said about not wanting me to come if I was going to cry and stopped.

"I'm not crying about you," I said. "I'm crying because I've lost my hearing."

"This can't go on," Carmen said.

"It can go on," I declared. "I overslept. How does that change anything? I'm moving into your room. I won't be dependent on those bells any more. You saw how I was awake the minute you touched me."

I stopped crying, but I felt heartbroken, about something more dreadful than failing to hear those bells. What had I been dreaming? If one sleeping pill did that, would a handful be enough or be too many for Carmen? But failure to hear those bells haunted me for a long time. For years I would sit up in bed thinking that I had heard bells ringing; and that I had failed to do whatever it was they had summoned me for.

I moved into Carmen's room that very night and there all was much better. So long as I had a room of my own, I kept the pretense of having a life of my own. It was like trying to reconcile marriage and virginity. One or the other had to be given up. In a deathwatch, one must give up one's own life.

Carmen and I became partners in the life she was living—or ceasing to live—and we were happier;

sometimes almost truly happy. No more bells; the touch of Carmen's hand or even a change in the rhythm of her breathing awakened me.

We were planners, technicians, plotters, true. And the goal of our planning was death. But we had all the verve of generals planning a battle: death would occur, certainly; the battle could not be waged without that, but the battle plan itself should consider all possibilities of setbacks and miscarriages and reverses. The battle itself should not fail because of inadequate planning.

"Mama wished she had saved Grandma two months of agony?" Carmen asked again.

"Yes. She did."

"Why didn't she?"

"I told you. Short of the butcher knife or holding Grandma in the horse trough, she didn't have any way. Besides, fifty years have changed people's thinking."

"It hasn't been fifty."

"Forty, then. People thought then that God wanted human beings to suffer. They enjoyed seeing suffering and violence themselves more than they realized, and they thought that God did, too. They made laws to see that no one had an easy way out of his pain."

"The laws haven't changed."

"O.K. We're going to be lawbreakers."

"It's exciting, isn't it?" asked Carmen without irony.

That was the last word I'd expect a woman dying of an inoperable cancer to use. But there had been a turnabout by making death Carmen's choice; by making its manner and time her choice, not fate's. It was now an act of defiance. She was no longer the victim, the hostage. She felt some of the high of a robber. She would take, instead of be stolen from. She would escape with her treasure, not have it filched from her coin by coin.

There were, as in any battle campaign or bank robbery, a good many emergencies to anticipate and plan for.

First of all, Carmen wanted her doctor's, Dr. Alfred Munger's, opinion as to how many pills she would need. I had already, in my home town, made inquiries of doctors about this for her. She didn't have much confidence in the offhand knowledge of men who didn't know her condition; and I didn't have much confidence in any doctor, except Dr. Wehrly, who had provided the Sedobrol, encouraged Samantha's residence, introduced me to beer, and sent me home at last, not to die, but to live and write.

The doctors I had talked with swallowed at once *my* story that the information I needed was for a story I was writing. But Carmen was not fictional. Her condition was a particular one, which Dr. Munger understood. He knew her resistance, her weakness, what could be expected after she had swallowed the pills.

"Talk to Dr. Munger," Carmen urged.

Talking to Dr. Munger was easy; but not about suicide; not about the death of his patient from an overdose of barbiturates. That wasn't the purpose of doctors. They could let patients die from failure to X-ray; failure to read a patch test; improper attempts at pneumothorax; premature dismissal from a san. And a patient as a result could have a nice, prolonged death. But a quick and easy death was not in the doctor's province. The patient, if he wanted that, had to hoodwink his doctor. So Dr. Munger and I were wary of each other. I did not want him sabotaging Carmen's plan, sending her off to a hospital where by tubes that drained and tubes that fed she could live indefinitely, no longer Carmen, but the patient dying in Room Number 37. And he, a Chicano, a doctor by means of brains and hard-scrabble, clawing determination, did not want this older sister involving him in any unethical practice.

I liked the man; first of all, because I liked his looks. He was no pink and white, plump bedside fellow who had become a doctor in order to gain a captive audience to talk to; an audience helpless under the sheets and propped up on three pillows. He was a real man, tall and gaunt, no bedside pundit and pill-pusher.

He treated Carmen as she was accustomed to being treated: as an attractive woman. If he saw inside her colon that tumor daily ballooning, he gave no sign of it. And Carmen, as she was still capable of doing, responded to him with the vitality that the female hormone produces in the presence of the male.

When my husband, on a hunting trip to Colorado, had phoned to talk with Carmen and me, her "Hello, Harry Maxwell" was so vibrant that Max, speaking to me afterwards, said, "She's a lot better, isn't she?"

She wasn't. But for three minutes Carmen had been what she had always been with a male: vivacious, pleased to be remembered, and responsive.

My first conversations with Dr. Munger were wholly practical. How often should shots be given?

The uses of Percodan and Parnate? Chats about his rose garden, his children, and the unusual rains that had given the valley its unseasonable spring.

Because of her experience with the cats, Carmen was determined to find from Munger whether or not an overdose of barbiturates would cause her stomach to reject them. Too few, she was afraid, would do no more than make her sleep for a couple of days. Would too many be even more useless?

I didn't know of any very good way to approach this question indirectly. Dr. Munger had fallen into the habit of sitting briefly at the breakfast bar in the kitchen after he left Carmen and drinking a cup of tea with me.

"Till noon, I drink coffee," he said. "After lunch it's tea. It's like turning the hourglass over. Day grows. Day dwindles. Tea strikes me as something to taper off on."

"There's the same amount of caffeine in both," I told him.

"Do tell," said the doctor.

Of course he knew that. Why did I say it?

"It's all in the mind?"

"Tea has an afternoon look."

"The English don't think so."

"You are a compendium of knowledge," said Dr. Munger, not admiringly, I was afraid. But he had given me an opening.

"How much longer do you think Carmen has?"

"She has a strong constitution. The tumor grows. But kidneys, heart, stomach, liver, function as well as ever. The bowels can't, of course. But enemas take care of that, at present. There might be months left."

"Suffering as much as she does now?"

"What she suffers now is a pinprick to what will come later."

"She doesn't plan for there to be a later."

"They never do at this stage. 'I will put an end to it.' At this stage they think so. But we are born with an instinct to survive, not to die. They postpone. Life is all we've got. Not a thing more. It takes a lot of courage to throw away all you've got. Even if it's only one rational hour a day. So they postpone until they're too weak and too drugged to do what is necessary."

"Not Carmen."

"You needn't worry about her. I've been through this a hundred times. You haven't."

"I'm not worrying about her. I know what she will do. I'm worrying about you. What you will do."

"I will certainly try to keep her alive as long as possible. You know that."

"I know that. That's what I'm afraid of."

Dr. Munger had a habit of lifting his hand toward his beard, then withdrawing his hand without touching it. He had told himself that he didn't care for men who fiddled with their beards; and before that could happen, his mind lowered his forgetful hand. He had dark eyes, of course. Not Jesus eyes; not Pancho Villa eyes, agate-hard, gringo-hating. But open, as dark eyes often are not. Or eyes of any color. How many conversations have you had with robots? Plenty of words said, but never an invitation inside to where the speaker lives. Or perhaps few do live inside. Life is gone, though speech lingers on in the mouths of these well-taught robots.

Dr. Munger let you inside. Possibly there was

nothing else he could do: the man inside was present in his words. He could keep his hand from his beard, but not his soul from his conversation—whatever his words were.

There were some tea leaves at the bottom of the cup and he chewed them like grass or tobacco. "It's such a damned lonely disease," he said.

I did not ask if lonelier than other diseases. It's a flaw of mine not to be able to ask questions, even when, as in this case, a question was perhaps expected. What I take to be diffidence is perhaps regarded by others as disinterest.

I've wondered silently about the loneliness of death by cancer. The slow, certain, measured tread.

Dr. Munger swallowed his tea leaves.

"You shouldn't let Carmen become so emotionally dependent on you."

I didn't question this either. "I'll be here for as long as she wants me."

Then I went back to the crucial question. "Carmen is not going to postpone for months. Or one month, if death seems better than life. You base your belief on experience with other patients. I base mine on experience with Carmen. She will not wait."

"What is the escape plan? A gun? A knife?"

"Sleeping pills," I said.

"I never gave her any."

"I know you didn't. Professionally, you're not involved in this. We've been careful about that."

"It takes a lot."

"She's got a lot. How many? For a woman in her condition?"

"Three," said Dr. Munger.

I was struck with speechlessness again. Three! I

could take three at 10:00 P.M. and be wide-awake at 3:00 A.m. Three might've been right for that cat, but not Carmen.

Dr. Munger walked to the kitchen door, then came back to me.

"How many times have you been out of this house since you came here?"

"I walk up and down the road at night."

"I mean clear away from here. Out of the valley?"

"Not once."

"You ought to."

"I'm O.K., Doctor. I can still count."

Then, into his eyes, where he lived, I saw Dr. Munger say, Don't ask me questions I can't answer. Don't.

But it was to Carmen's need, not his, that I was vowed.

I had to wait until the day nurse left before I was able to talk to Carmen.

"What did he say?"

"Three."

"What?"

"He'll loosen up. I know he will. He doesn't trust me yet. You can't expect a man to risk a practice he's worked twenty-five years to set up, to throw it away on a little inside knowledge to a busybody relative."

"Did he call you that?"

"No, of course not. He just said that he knows more about patients than I do. And I said that I knew more about you than he did. And there it was a standoff. He chews tea leaves."

"Marijuana?"

"No, no. Constant Comment."

"Dry?"

"No, from the bottom of his cup."

"I like him."

"So do I. But you're going to have to trust what the doctors up north told me. And what I've read in doctors' books."

"You think about death as long as you hope to escape from it," Jules Renard wrote in his journal.

Carmen, who did not hope to escape from it, thought instead of her plan to escape agony.

"People who don't know anything about it," she said, "talk about 'death with dignity.' What the hell do I care about death with dignity? Others might like to be spared the sight of something undignified. If I don't feel it, what do I care? I've never done a thing so far in my life in order to appear dignified. I'm not going to start now. 'That wouldn't be very dignified.' Did that ever stop you from doing something you wanted to do?"

"If it did, I never thought about it in that way."

"It makes me sick—stupid thing for me to be saying—these prosuicide people claiming that *that's* what they want to provide for us. Dignity! It's not dignified to be born, a bloody, messy business. It's not dignified to get a baby. It's not even very dignified to eat. And they should see how I manage a bowel movement now. No. If there was some way they wouldn't have to be mixed up with it or witness it, they wouldn't give a damn about dying with dignity. And I don't give a damn. If I'm not conscious, what do I care? Nobody unless he's crazy wants pain. Nobody unless he's crazy thinks about dignity."

Carmen and I had two subjects of conversation.

Three, counting the progress, if any, I was making with Dr. Munger. First, the pitfalls we might encounter with the pills. Second, and once in a while, old times. Carmen didn't have the dread I had had of remembering my lost past. Partly, perhaps, because her disease was not one she had brought upon herself; she had at least not wheezed around with broken-down lungs signaling trouble for two or three years. The blows fate gives us we suffer from. But they don't fill us with remorse. The blows we give ourselves, "Except for my own stupidity I wouldn't be here now," those we explore and re-explore. "Oh, God, what made me do that?" Or fail to do that? That torture the self-hurt must try to avoid. Carmen wasn't self-hurt. She had been careful about checkups, had not postponed her operation. She suffered now from what had been as much a part of her genes as the frame that prompted beach-watchers to want to take her picture. She was living the life the body she had been given had destined her for.

Nothing more could be done about that. Something could be done about the death of that body.

First of all, and without any advice from Dr. Munger, we knew that the pills might not cause death in twelve hours. The plan for taking them was already laid out: at her bedtime, the big dose for her; a few for me. We would go to sleep together. I would wake up. She wouldn't.

But would she still be asleep in her drugged sleep when Goldie Thorsen, the day nurse, arrived at 8:00 A.M? Goldie was all nurse, while on duty, anyway. She, like Dr. Munger, had achieved her profession in spite of obstacles and would not willingly be separated from it.

Rowena Martin, the cleaning lady, came three afternoons a week. Carmen and I both looked forward to her arrival. In the first place, she treated Carmen as if she were well—taking a siesta, perhaps, after a late night out. Secondly, she was a wonderful storyteller, and her stories had to do with her raunchy husband, a fellow she pretended to disdain but obviously took pride in. His exploits were sexual and she was the victim, as she told the story. But she was the victim and the reporter of her victimization too often to give much credence to her pretended role as martyr.

LeRoy (do all sexually aggressive men have an *o* in their names?) was a man who liked diversity in his sexual activities, but not the effort required to find other partners. Whether or not this was the real reason, he was faithful to Rowena. He put excitement into continued copulation with the same woman by varying the circumstances and locale. Sometimes they were LeRoy and Rowena, husband and wife of twenty years' standing (and frequently lying, according to Rowena's stories). Sometimes LeRoy assumed the role of a complete stranger overcome by Rowena's sex appeal.

Once he pretended to be a Mexican wetback, hiding from police, and took a Rowena, feigning fear, at the bottom of an irrigation ditch, hidden from view by overhanging palm fronds.

"He talked Spanish to me like a native," Rowena said admiringly.

Once he came to the home of one employer, undressed her in the rumpus room, and did the job on the billiard table.

"My heart was in my mouth for fear someone would come in," admitted Rowena, "but LeRoy don't know the word 'fear.' "

We waited with something between apprehension and anticipation for LeRoy to put in an appearance with Rowena some afternoon under the oleander bushes in the patio.

"Will we watch?" I asked Carmen.

"You bet." Carmen laughed. She could still laugh. "I'm never going to get to go to one of those new pornographic movies. Pornography will have to come to me."

Rowena was our entertainment, not one of our problems. Goldie was a problem. She was efficient, energetic, and as dedicated to saving life as Dr. Munger. We would miss Rowena, but I could do her work if necessary. Goldie's work was beyond me: the daily enemas Carmen now had to have as she lay in a tub of warm water did not faze Goldie.

"Goldie's not going to break her vows or give up nursing just because they interfere with some plan of ours," Carmen said. "We've got to figure out what to do when she arrives some morning and finds me in a coma. She'll call an ambulance. Or pump out my stomach herself. She could do it all right."

I figured out something. Unexpected days off for Goldie. I would call her in the evening and say, "Carmen's feeling so well I'll take over tomorrow. Same pay for you, of course."

With this practice established, we were prepared for the night when Carmen's hand, of its own will, reached for the pills.

So Goldie's possible threat was taken care of, and we could enjoy her efficiency and good humor. She laughed as much as we did at Rowena's stories.

When Goldie left each day at four, I would go into Carmen's room. I tried to stay away when Goldie was there. If I didn't, I spent the energy needed for the night in talk.

"Well, guess where Rowena got it this time?" Carmen would ask.

There were very few places left. "Not a supermarket?"

"He wasn't crazy."

"No? What's your word for it?"

"Innovative."

"You sound like a school administrator."

"Now, Harry Maxwell doesn't use words like that."

"His buddies do."

"In the woman's side of a rest-area comfort station."

"In one of the cubicles?"

"Yes."

"And you say he wasn't crazy?"

"Even Rowena thought that was exciting."

"I think she's started making up these stories to entertain you."

"Maybe."

There were other matters that could not be taken care of by paying off Goldie. "The chief thing," Carmen said, "is not to get back into the hospital, ever. They're determined to keep you alive there. Do you know what they call suicide in a hospital? Autoeuthanasia. Yes, they do. It's like some kind of a game for the doctors. The winner is the one who keeps his patient alive longest. Oh, I know. I've been there."

"You don't have to worry about being there again."

"Munger could send me."

"He won't."

"What if some relatives arrive. You know the ones I'm talking about. I've taken the pills and they call the hospital. What're you going to do?"

"I'm going to carry you out through your doorway down your patio to the garage and put you in the car and drive off."

"While they're in the living room?"

"Right. Then I'm going to drive up to that little hidden canyon no one knows about except you and me."

"And let the pills take their course?"

"I'll not depend on that. I have a little piece of hose. I'll attach it to the exhaust and leave the car running."

"Where'll you be?"

"Outside, wrapped up in a blanket."

"When it's finished, you'll drive back down?"

"Yes."

"You'll land yourself in jail, that's what you'll do."

"It won't come to that. You know that. All I'm going to have to do to anyone who insists on seeing you is to put my foot down, say you've just had a shot and have gone off to sleep and that I will not permit you to be disturbed. There won't be any trip to the canyon. There won't need to be. But there *could* be if the worst came to the worst."

"Some people would say it had."

"There's a lot some people don't know."

• • •

I don't know just when or why the change came over Dr. Munger. We had been doctor, patient (a patient he liked), and patient's relative (usually an additional worry for a doctor). Suddenly, no not suddenly, finally, we were three human beings; one dying, two involved with making that death as easy as possible.

He was the one of the three who had the most to lose. Carmen's life was already lost to her. She wanted to avoid agony. I had almost nothing to lose. But Dr. Munger, if he were suspected of co-operating in any way with the premature abatement of Carmen's agony, would lose his life's meaning, his pride, his livelihood, all that he had worked the first half of his life to attain.

He was not, himself, going to lift a finger to end a patient's life. He *could* not do such a thing. But I did not believe that he would lift a finger to prevent his patient from choosing the *time* of her death. Carmen was still worried about the failure of her attempt at euthanasia with her mother cat. Would her stomach reject too many pills? She wanted this information, not in general terms, but specifically about her, from her own doctor.

I dreaded asking Dr. Munger. It wasn't a subject he enjoyed talking about, or even felt he *could* talk about. O.K. One of us was dying and the other two could put up with a little conversational discomfort.

Dr. Munger had his usual cup of tea at the breakfast bar in the kitchen before leaving. This is not a place at which I like either to eat or drink. The view is of kitchen mechanisms—the ice-producing machine, the dishwashing machine, the egg- and batter-beating machine. Eating in such a place is like having dinner on the Santa Fe Chief with the engineer in his cab.

It *was* convenient for the five-minute chats I had with Dr. Munger before he left; it was as near to an operating room in appearance as you could get. It did not in any way suggest the pleasures of conversation; nor did Dr. Munger want these. His calls were not sociable, not even therapeutic. They were ritual. A patient of his was dying, and a doctor, though he can no longer be of medical help, does not abandon the living.

I had asked him late one afternoon if he would like a drink. I meant a Martini or Canadian whiskey on the rocks; not tea. He looked as if he needed something more relaxing than Constant Comment, whatever those chewed leaves did for him.

He turned the invitation down quickly, firmly, and with some asperity. His refusal was a rebuke. "I have a drink with my wife when I get home," he said. *Then,* he would relax; *then* he would put the death he could not prevent, and the agony he was not permitted to lessen, completely, as far out of his mind as possible. But not here. Not the happy hour in the parlor while the patient three doors away had nothing more to sustain her than Percodan. And Carmen, before her sickness, had been a girl who relished a drink as the sand hills turned pink. He knew that. He also knew that his wife looked forward to her role as the source of the drink and as the set designer who would provide him with the change in scenery he needed from hospital beds and urinals and the patients whose eyes daily asked him questions they did not want him to answer.

I understood all this the minute he turned down my invitation, and cursed myself for the insensitivity that had permitted me to make it.

But this afternoon, following his talk with Carmen, he picked up his cup of tea and said, "There's going to be a sunset to end all sunsets. Why don't we drink our tea in the living room where we can see it?"

I wasn't drinking anything. But I certainly preferred sitting in Carmen's elegant blue-and-white living room, watching the desert's dry bones brought to life by the setting sun, to watching the unchanging surfaces of kitchen appliances.

Since Dr. Munger had made of his own free will this move away from the clinical atmosphere of the kitchen, I got up my courage to ask him my questions. I told him about the cats.

"Will too many pills cause nausea?"

Dr. Munger had reached the tea-leaf-chewing stage. He smiled a little around the leaves.

"I'm not a vet."

He was not going to answer that question. I asked another.

"How long does she have?"

"I don't know. Organically, nothing is wrong with her. At the moment she just has this mechanical obstruction. It is painful. Without the enemas, she could not live. The tumor will grow. Will rip and tear. It is like a cannon ball working its way through her. Eventually, she will bleed and fester and putrefy."

"She will not wait for that."

Munger said once more, "That is what many patients believe at this stage."

There was no point in telling him, "Carmen is different."

"In two weeks I am going on a vacation," he said. "I'll be gone for two weeks. Dr. Duncan will fill in for me. In any crisis, he'll see that she gets back to the hospital. You understand that? Doctors are like soldiers. The soldier cannot ask on the battlefield, 'Would this man I face be better left to live?' Doctors cannot ask, 'Would this patient be better off dead?' The soldier is sworn to take life. We are sworn to preserve it. You understand?"

"I understand."

The sun had gone down. The sunset we had faced and had left the kitchen to see had happened without our noticing it. It was not yet black night, but it was gray twilight.

Bill came in from his office and, having seen Dr. Munger's car in the driveway, went down the back hallway to Carmen's room expecting to find him there. Carmen no doubt told him why I was talking with the doctor. Bill did not leave her room till Munger left.

Munger had either finished his tea leaves, or the light had grown too dim for me to see him distinctly.

The occasion was so solemn, the hour so bleak, sun gone, stars not yet out, that I, as I do too often when an occasion is solemn or sad, played the clown; thinking, I suppose, to bring a little sunshine into the darkness.

"Borges said," I told the doctor,

> *"the proofs of death are statistics*
> *and everyone runs the risk*
> *of being the first immortal."*

I did not expect laughter. Still less did I expect Spanish.

"Las pruebas de la muerte son estadisticas
y nadie hay que no corra el albur
de ser el primer immortal."

"That is the Spanish," I said, surprised.

"I spoke no other language till I was seven."

"You know Borges."

"How be Spanish-speaking and not?"

"Be poor," I said. "Have no books. Do not care for poetry."

"I was not poor. We had books. I care."

He stood and I switched on the lights.

"We understand each other?" he asked.

"Yes."

I went to Carmen's room. Bill had already left.

"He is with us," I said to Carmen. This, I believed, was the truth. the next sentence was a lie. "He answered all of your questions."

"To fall in love is to create a religion that has a fallible God." That is Borges again.

And a man somewhat like him, Camus, said, ". . . there is only misfortune in not being loved; there is misery in not loving."

Carmen and I, now that the planning was taken care of, tickets bought, destination accepted, talked more and more of the past. It was not a past we had shared when we had been of an age to remember much about it. True, she had been put into my four-year-old arms on the morning she was born. I could remember that; she couldn't. We were never in school together. She was not only "little sister." She was a little sister who did not enjoy the activities I did. No walks in the hills for her, no basketball games, no debating teams. But there was enough in me that enjoyed clothes and shopping and boys, and enough in her that could read poetry, to keep us close.

And there had always been the two qualities, one in me, one in her, that bound us. Or bound me. Her beauty; and the fact that of my family, she alone understood the fever and excitement of words: reading them, writing them. There was one other quality: the motherliness in me, augmented by my being the oldest in a family with an ailing mother; and of being childless after marriage.

This mothering lasted longer than healthy mothering should; and Carmen talked of its termination in what she called "The Summer the Worm Turned."

Two years before her illness, Carmen and I had gone to Europe together: her first trip, my fifth or sixth. As Mother, as the more experienced traveler, I had slept in upper berths, had occupied the back seat of rented cars while Carmen enjoyed the better view in the front seat by the driver, had downed hard liquor because Carmen didn't like wine.

Then one day, after boarding the hair-raising fast train from Madrid to San Sebastian, Carmen said to me, "Change seats."

"Why?" I asked, truly wondering.

"I am facing backwards."

It was then the worm turned, and discovered in turning that it had lived with Carmen a wormlike life. Not a life Carmen necessarily admired; but one adapts oneself to the nature of the creature with whom one finds oneself coupled.

"Why do you want to face forward?"

"The view is better."

"O.K. For half of the journey, I'll stay where I am. Then I'll trade with you."

Carmen laughed then and she laughed now, asking me, "Do you remember the summer the worm turned?"

"You don't like worms. Why didn't you tell me?"

"You suited me as you were. Why try to change you?"

"You would have liked me better."

"Maybe. Maybe not. I like being waited on."

This was the idea that had struck me the summer the worm turned. It was not the best of summers, discovering that I had a fallible God. But that discovery, with the rift it brought, disappeared as I learned the truth of Carmen's insight: "not being loved is a misfortune; not loving is misery." The worm turned again, again the loving continued, the misery abated.

Natures and temperaments differ and are largely unchangeable. It takes more grace to receive than to give, a grace I have always been short on. I remember with pleasure the gifts I have made. Those I have received I forget. It may be more blessed to give than to receive, but there is more grace in receiving than giving. When you receive, whom do you love and praise? The giver. When you give, the same holds true. Carmen's grace in receiving had made me a lifelong lover of myself, happy in bringing happiness, I believed, to her.

Grace was one of the few able to find joy in both giving and receiving. Her thank-you letters made jewels out of the trinkets she received from me. She gave of her whole self. She did not send presents gift-wrapped from Gump's or I. Magnin's. She herself was the package, bleeding, aching, and finally dying to help those she loved.

Carmen and I had many differences. She did not intend that one personal item of hers should be left in her home when she left it for the last time.

"I thought it was terrible," she said, "Mama's things still in the house years after her death."

I had thought it nice. Nightgowns in the same drawers. Pictures still pasted on the inside of cupboard doors. Recipe books where they had always been. Harmonica near at hand on top of the piano.

This was not the way Carmen intended it to be for her. She was not a Pharaoh, who planned to take with her war horses and concubines and golden bowls of fruit. Not one personal thing of hers would be left in the house at her death if she could help it.

She disposed of her clothes, and she had many, in a way I could never have managed. She was small, slim-waisted, high-busted; her clothes would not fit the matrons of her own age who were her friends. They went, instead, to her friends' daughters; and the daughters came in to try them on so that each could receive what best suited her. And receive also Carmen's advice as to where a scarf should be added, a hem let down, a belt removed.

If this gift-giving seemed ghoulish to me, it did not appear to seem so to either Carmen or the girls. They were artists working together with the materials at hand to achieve something beautiful. And beauty created can outlive death. I stayed out of the room for the most part while these fittings went on. But on the time or two I came in, I was not noticed.

The atmosphere was that of the concentrated attention of any fitting room: Carmen sitting up in bed, the girl in front of a full-length mirror, harkening to the advice she was receiving about alterations.

Carmen's clothes could not be altered to fit me. But mine could, Carmen thought, be improved. I had come straight to her house from Oxford, Ohio; and what was being worn in October in Oxford, Ohio, was quite unlike the garb of Palm Springs and its environs at that date.

I went, I bought. I returned and was instructed by Carmen, "You can do better than that." She was right. I could. I went again and returned with better.

There were times when for an hour Carmen could sustain her role, or sometimes it was Grace's, in a scene from the past.

I was sitting in the late afternoon on a sun-tan cot back of the house when Carmen came to the patio gate and called, "Whoo-whoo." Carmen had never whoo-whooed in her life. She was Grace calling to her girls, still children perhaps, to come home. She had remembered something and she wanted to make the memory live for me. She did.

She came to the sunroom off her patio in the middle of a Saturday afternoon, in white slacks with red bandanna halter.

"How about a hand of bridge?" she said to me, Goldie, and Bill.

This was not tuberculosis, a disease in which you urged the patient to take care, to go slow, to husband his strength. He could do whatever he felt up to doing; and the more he felt up to doing, the happier he made those with him.

Carmen and Bill were practiced players, members

of bridge clubs and winners of trophies. They divided Goldie and me between them as equal handicaps, Carmen getting Goldie, Bill, me.

I liked games, had a winner's instincts, but was ignorant of the conventional signals given by a partner's play. Besides, I wanted Carmen to win. My desire made me clumsy, for which I was glad. Carmen didn't want the gift of a game. She wanted to win it. She did. It was time past regained.

Carmen did not talk of death; lament her fate; wonder about the possibility of life after death; cry for more; mourn over past mistakes. She was a stoic. The Indian who gave the slant to her eyes and the olive hue to her skin lived inside her at her dying. Her talk, and preparations, were for a journey. She expected to be sleeping when the flight took off.

She did plan for those who would not be accompanying her. She had me call six of her nearest friends from the phone in her room, so she could hear what was being said, and ask them to invite Bill to dinner parties to which attractive women would be invited. Women who might succeed her.

"If he is left alone, Bill will mope and drink too much and die. He is no chaser. What he needs is some good woman who will make a play for him. You tell them that. You say that I told you to say it."

I did. But I cried while I was doing it. My back was to Carmen and she couldn't see the tears; and the hoarseness she took to be some hangover from my old chest troubles.

She planned her funeral: in the Quaker church a step down the road from the house in which she had been married and where Grace's funeral had taken place.

There were only two things she really cared about: the songs that were to be sung, and the complete closure of the casket during the ceremony.

"I will not be stared at when I can't stare back."

There was to be no "Shall We Gather at the River" or "In the Sweet By and By."

There would be two hymns, one surprising to me: "In the Garden."

"I love that song. I've always loved it. It makes me cry. 'Oh, He walks with me and He talks with me, and He tells me I am His own.' "

It was more like a love song than a hymn. (And a good many hymns are that.) And it told me something of life's lonesomeness. "He walks with me and He *talks* with me." This of a thirty-year-old dead two thousand years.

The second, also a hymn, was not surprising. "Now the day is over, night is creeping nigh, shadows of the evening fall across the sky."

Her third choice was her long-time favorite, perhaps never sung before at a funeral and as little appropriate perhaps as "Life Is Just a Bowl of Cherries" or "Cabaret." It was "Ah, Sweet Mystery of Life."

Life was that. A mystery. Who would deny it? And if sweet to Carmen, who was in any position to deny *that?*

"It's not very highbrow," Carmen said.

"The hell with that."

"I'm not very highbrow. Highbrows can walk out while it's sung if they want to."

"I'll announce it. It will be interesting to see who goes."

"There's one other thing. I wish I could be buried naked."

"Why?"

"It seems such a waste. Doll up when no one is going to see me. Wear something that'll only stick in the teeth of the worms."

"Wear something cotton. Or wool. That would be ecological. Like a leaf falling. Or a sheep dying. The worms wouldn't mind that."

Carmen laughed. "The worms! A sheep dying! Such talk at a deathbed. Give me a shot. I hurt."

"That talk didn't cause it?"

"No, it's half an hour past the time. I got so interested in my funeral I forgot my pain."

Rain, so scarce on the desert, and almost unheard of in October, came in October; and heavy enough so that sparse grass greened the October sand.

Carmen took it as a personal gift. She had been given two unexpected seasons: winter's rain and now spring's transient greenness.

"If suddenly we have summer heat, I'll think everything's been miniaturized for me. Each day a season. A month, two years. There are children like that, you know. I've read about them. They are ninety years old when they die at ten."

"That's just in their bodies."

"How do we know?"

We didn't, of course. I didn't even know what was in Carmen's mind. In my egotism and love I thought I had known what she was like and what she had wanted. I had too much backwoods, Quaker reticence to inquire.

No one in school or out had ever spoken to me of sex—nor had Carmen and I ever talked of it. I did not think of myself as a prude (one, somebody said, who renders unto God that which is Caesar's). But perhaps, prude or not, I gave that impression. Carmen and I, who had lived each with a man since youth, had never talked of these two men—or of any others. I doubt that it is possible to truly know another woman who cannot or will not speak to you of what she felt with her husband. But the time for that was past, long past.

The heat of summer did follow almost at once the faint premature greening of the dunes and hills. But summer heat was nothing unusual for October in the desert, the thermometer up to 110 degrees day after day. There was air conditioning in the house, and heat as heat meant nothing to us inside. But heat as an unexpected summer delighted Carmen.

Dr. Munger went off on his vacation. On the afternoon before he was to leave, he brought Carmen, as usual, flowers from his garden; to both of us he gave instructions about Dr. Duncan, who would substitute for him.

"He's a good man, but he'll be busy. He has his own patients. If your bladder stalls again, Carmen, you'll have to have him to insert a catheter. Don't trust Goldie for that. There might be other things. But I'm going to tell him he doesn't need to be dropping in every other day as I do."

"You're jealous," said Carmen.

"I won't deny it. So long, kiddo."

Munger bent over and kissed Carmen on the forehead. "Your sister knows *almost* everything I do by now. And what she doesn't know, Goldie does."

"Have a good time," Carmen said.

I went to the door with Dr. Munger.

"No tea," he said. "We're packing at home."

"So are we," I told him. "In a way."

"In case of any emergency, Dr. Duncan will get Carmen to the hospital. You understand that?"

"That isn't what Carmen wants."

"You know what Carmen wants. I know what Duncan will do."

"If I come down with acute appendicitis, you recommend Duncan?"

"He's a good doctor. I'm not sure you'd be a good patient."

"Doctors can't make choices like that, can they?"

"No. Worse luck."

The summer heat persisted. Carmen's closets were empty of her clothes. Every personal item of hers was packaged and labeled. She had drawn pictures of all the dishes of Grace's she had; under each picture she had written the dish's history. "A wedding present from Aunt Lib, Mama's mother's sister." "Gift on her sixteenth birthday from Enoch Miles, the hired man."

The objects that she and Bill had acquired were to remain with Bill. Such notebooks and engagement books as she had kept were put in a cardboard box for me. They were for the most part brief timetables; they revealed little.

She no longer took sleeping pills. "They give me terrible dreams." I did not ask her what the dreams were. Or perhaps that was a subterfuge. Perhaps she did not want, by taking a few pills now, to blunt the effect of the handful she would take later. She no longer took Parnate, an antidepression preparation she had.

"You take it," she said. "You need it more than I do."

I took a few. Powerful stuff. I no longer felt like a woman at the bedside of her dying sister, but like someone bathed in the sunshine of an unknown glory. It was too unreal to endure. I also put the Parnate away.

The enemas had now become major operations. Carmen's arms had become too riddled by the needle to be useful for further shots. The painkiller now went into the buttocks, larger doses and more frequent.

Carmen once asked me, "If it were you, would you end it now?"

I didn't know, of course, and I spoke more out of my desire not to lose her than out of any real knowledge of suicide. I, when ill, with far less vitality left than she, had not been able to get into that tub with the electric heater. But, then, I hadn't her courage.

"If I had one good hour a day, I'd think twice before throwing it away."

The best hour seemed to be that after dinner and before the heavy bedtime shot.

Unlike me, who had wanted to forget that a world outside my room existed, who could not bear to think of either the past or the future, Carmen continued to relish the present. She watched television; she could talk about the past, and the future that would still exist for Bill and her son. She relished them; relished them, that is, when the shot had not taken her away from consciousness as well as her pain.

A brother spent the day with her, and left without her having said more than a half-dozen words. What Munger said was perhaps true. The will to shorten the long-drawn-out process of dying weakened and died before death itself came.

The night after the brother's visit, I said, "You didn't say much today."

"I didn't have much to say."

Carmen and Grace were unlike in this. Grace had been determined to be a good hostess if it killed her.

Carmen was her father's daughter, more Comanche than Welsh-Irish. If her guests had come with news, let them speak. She would listen. If not, she would be silent. Their visit was none of her doing. Grace thought that guests should be entertained.

Carmen disliked Grace's habit of bewailing to us the arrival of guests, then of convincing them that they were the very persons for whose presence she had been longing. Grace followed them to the car so that they left believing that they had brought sunshine into a lonely woman's day. That's what *they* thought. That's the role that Grace, the actress, with no other stage than the family living room, played.

"They should have seen her tottering into the house hunting the Musterole and aspirin," Carmen said disapprovingly.

And denouncing in terms still dramatic persons who outstayed their welcome; persons who were unable to take a hint.

"Hint?" Carmen would ask me. "What hint?

Mama convinced them that they were the joy of her life. They had never in their lives had such success as conversationalists. And all the time, Mama praying to God they would leave. It was hypocritical."

That wasn't my word for it. It was foolish, and in the long run suicidal. But this was frontier hospitality, in the practice of which Grace had been brought up. The guest was made welcome. And Grace, the less she felt like it, the further she pushed herself (out of her conviction of mean-heartedness) into the role. Understatement wasn't part of her nature.

"The more she wished them gone, the more she felt the need to hide it. She overcorrected a bit."

"It wasn't honest."

Carmen was honest in a way denied Grace and me. When she was tired of a visitor, her part in the conversation dwindled to nothing. The presence of another human being did not constitute for her a drama in which there were lines she had to speak. Monologuists blossomed when she was bored.

She had the grace to accept—but only what she wanted. At Christmastime, the gifts she had received but had no desire for were left at Grace's, where those yearly celebrations were held. She made no apologies; she attempted to hoodwink no one. What she didn't want she left behind.

I have to this day four china chickens of colors most unpoultrylike and glazes most blinding. Two were bought for me, two for Carmen. I took Carmen's two, scorned by her, home with me for Grace's sake.

"Just what I've always wanted," I told Grace, who had, I knew, shopped for them, varicose veins tightly bound with elastic bandages, and stomach awash with a solution of bicarbonate of soda swallowed to counteract the acid of aspirin.

Grace brightened a little, I thought, hearing my enthusiasm. Maybe not. It takes a thief to catch a thief.

Carmen lived a more straightforward life than Grace and I. Perhaps we all did the best we could with the temperaments we had. Carmen could sleep, if she didn't feel like talking, though a visitor sat by her bedside. Grace and I couldn't.

But she wasn't as direct as this may make her seem. Goldie and I between us cooked dinner. Goldie started things; I finished and served. Bill sometimes with us, often not. He had many meetings to attend, and except for his own poor health, a political future ahead of him.

One night the main dish was roast lamb.

"I've always loved lamb and pork," Carmen said, "and almost never had any."

"Why not?"

"The men in my family wanted beef. Chicken, once in a while. But beef: steaks, roasts, stews. Pitiful, isn't it? Have to die to get what you want to eat. And vulgar to have this appetite now. But Goldie tears me apart. I swear when she's through with me I'm emptier than I was when I was born."

I knew what those ordeals were like. Goldie had told me. Agony in the morning, and as a reward a bite or two of the food you had been denied all your life for dinner. Was it life's irony or pity that made my eyes fill?

They did, because Carmen said, "You promised me none of that."

"I'm keeping my promise. But do you know what I did each morning when I was sick at home? Cried. Cried my eyes out. The minute I opened them."

"What would you have done if Mama had come in each morning crying *her* eyes out over poor little you?" Carmen asked.

"I don't know. I never saw her cry. Did you?"

"No."

"It's odd that she didn't. Moderation wasn't her habit. Not in getting mad. Or laughing. Or slapping us if she felt we needed it."

"Or if it would relieve her feelings."

"I suppose so," I admitted.

"Papa could cry, though."

It was the truth. Eldo, had he not lived with whites for so long and picked up their ways, might, amongst his mother's people, have been called Old Rain in the Face. No sobbing, no flinching at pain of his own. I saw him pull out one of his own teeth with a pair of pliers. But let a mishap befall Grace and the tears fell.

"Mama had that hard Irish streak," said the red man's daughter.

Maybe she did. For the months I was in her care, she needed it.

Carmen usually looked at television after dinner. On the night we had leg of lamb, she was in pain and I gave her her shot early. Then I, instead of reading or watching TV with her, went for a walk.

The moon, a few nights past full, was already up. It had bleached the rose out of the sky above Tahquitz. The days were still warm, in the nineties, but once the sun went down we were a planet without heat. There were more quail and cactus and sagebrush and centipedes on the road on which Carmen's house stood than there were other houses. The few houses that were there were opened to the night air—and to onlookers. In them people trod the usual evening path toward death. In them, all undoubtedly knew what was happening in the back bedroom of the house at the top of the road.

There is a remedy or there is none. What you can't help, forget. With such sayings we insulate ourselves.

I didn't hold their Martinis or tape recorders or bridge games against them. I didn't cry every morning for Carmen as I had cried for myself. I didn't, as Grace had done for me, set myself the task of reversing fate, of bending it to my will.

True, I was saying good-by to the silvered sagebrush leaves and their smell, half of the desert and half of Thanksgiving turkeys, to the desert's sundown bone-piercing cold and the dry slither across sand of lizard and mouse, for Carmen. But I was present, too; the half that mourned didn't completely negate the half that enjoyed—and who had earlier been reprieved.

Carmen was awake when I got home, not hurting.

"What was it like?" she asked of my walk.

But when she saw that I was trying to make a present to her of the world she had lost, she turned it down as flatly as she had the Christmas chickens.

"I said good-by to all that last summer. It wasn't easy, but I did it and it's over. All I'm saying good-by to now is pain. You get this cannon ball in your gut and see how much you care about moonlight on the sand dunes."

What she said was true. I was feeling what she had felt last summer: the world's beauty, lost. I was half a year behind Carmen. The world lost, yes; but in trying to share that past beauty with her, I was reminding her of past sorrow. But the cannon ball in the gut with its time fuse set to detonate, no. I could not share that. Death, departure, extinction: the generalizations, yes. But the cannon ball never. Moonlight on the sand dunes, *sans* cannon ball; *that* was what I had been enjoying and had brought back to Carmen. As well give a dancing girl to a eunuch. I was an idiot.

Carmen didn't go on with that. She returned to a past before sickness came. As the dosages increased, she went there more and more often.

"Do you remember Sedbergh?" she asked.

"I remember."

"You came back for me, didn't you?"

"Yes."

That was the summer the worm turned, the summer when I was trying to follow with Carmen a path dictated by my head. Harry Maxwell had joined us in England and we had a car. Try to act like a grown woman with a husband, not a quasi mother with a baby sister, I reminded myself. So when my husband had said, "Let's drive to that little village back in the hills," I had said, "Yes," not my usual, "Wait until I see if Carmen wants to go." The village, name forgotten, high on the Yorkshire moors, with a stream in the swale below it, and sheep and stone walls on the broad-shouldered down that rose above it, was the loveliest artifact human beings have ever put together. I didn't want to see a single curve or a color of it with Carmen sitting back in her room reading old copies of *The New Statesman.*

"We have to get Carmen," I said.

Max, fond of Carmen in no wormlike way, said, "You shouldn't have gone off and left her."

She was standing, when we returned, at the window in her first-floor room, looking from the window down the road that we had traveled. She was downstairs with her coat on by the time the car was turned about for the return trip.

"You came back for me."

The truth was, it was nothing without her. I didn't say it. She could swallow only so much worminess, made vocal.

The days when Rowena's stories no longer interested her, and her own speech was sparse, grew more frequent. The sessions in the bathroom were longer. There were groans that could not be repressed carried to my bedroom down the hall.

Goldie said to me what Carmen had said to me earlier, "This can't go on." But she added, "We should call the doctor."

"Wait until Munger gets back. It's only another day or two."

"You don't know what's coming out now."

"I know I don't, Goldie. But she has her ups and downs. She eats one of your custards in the night, smokes, talks of old times. You see the worst. A person could come in in the night and think that the worst she had was a bad cold."

Goldie nodded. "For her it's like being killed in a torture chamber. If the growth was on her liver or her lungs or her pancreas, they couldn't function and she'd be sick. The colon is nothing but a drain-pipe. It doesn't furnish her with anything she needs to live. It's the damnedest way to die."

"She still has her good moments."

"After I've finished with her, maybe."

"You need a day off, Goldie."

The twenty-fifth of October was a day a good deal like those that had gone before it. The late afternoon, when I sat outside, was hazy and all but colorless. Many birds were cheeping. No other sounds except cars far away on the highway; and in Carmen's room, the TV blaring while she and Goldie slept. She wanted to sleep this afternoon if she could do so, in order to stay awake to watch *The Bad Seed* on TV that night. In some way, I believed she thought that the cancer was the bad seed she carried.

Munger came back from his vacation. I think that he was surprised to find Carmen still alive. After he had finished talking with her, he took a seat once again on a stool facing the kitchen appliances. No tea, though.

"Carmen let me examine her this afternoon."

I looked at him inquiringly.

"Before this she has refused."

I still couldn't ask.

"The tumor is two to three times as large as it was six months ago. Her physical condition otherwise hasn't deteriorated. If anything, she's heavier and stronger than she was. She has a low pain threshold. She is getting opiates much stronger than I would give to anyone else. She will not die of cancer. She will die of lung congestion or heart stoppage."

"When?"

"With her constitution, it might be a year. Alive, but not really living."

"If you were in Carmen's shoes, what would you do?"

"If I were in my own shoes, I know what I'd do."

"What?"

"I wouldn't ask a doctor a question like that."

I asked no more questions.

"Physically, she is if anything stronger. Mentally and emotionally, she has deteriorated. As the opiates increase, the memory, the will, the determination will grow weaker. I told you this earlier."

He had. I let it go at that, without arguing that Carmen was different.

The last two days had been particularly rough for Carmen. Always before, the painkillers had, for the time being, killed her pain. Now, though the shots lessened the pain, she was never completely at ease. Her afternoon's sleep while she and Goldie were deaf to the soap operas had given her her most complete comfort for several days.

I had, with Goldie's help, prepared that evening an old-time, back-East, cooked-by-Grace dinner. Pork chops, which Carmen liked, brown milk gravy, sweet potatoes, sliced tomatoes, apple sauce, coconut cream pie. Not a meal for an invalid; but Carmen was less an invalid than a maiden tied to the tracks by a villain and waiting for the Cannonball Express to roar by.

She ate with relish, had a shot, watched *The Bad Seed,* then said, "Do I have to wait any longer?"

"No. You do not have to wait any longer."

The next hour and a half passed like a gale, silent and black. There were a hundred things, it appeared, that in our months of planning we had forgotten—or I had forgotten. Casseroles and pudding dishes to be returned to their owners. Letter to Eldo to be stamped and addressed. Check on funeral arrangements: three hymns and Baldwin's words for Rufus at *his* funeral. Did I have it copied, and if the minister wouldn't read it, would I? I would. She had on her fanciest and prettiest teal-blue silk nightgown. Had she known when putting it on, after her session with Goldie that morning, the reason for donning it?

She wanted silk panties, too. She knew that she would be examined, stripped, too, of course, but she didn't, probably couldn't, yet think of herself as a corpse. When it was all over and the doctor was called, he would find a woman decently covered and with her pants on.

But before she pulled them on, and she could do this—she had never been tied to bedpans as I had, or too week to stand—she asked for the bottle of Givenchy's L'Interdit that I had brought her from Paris when I came home from my script-writing job. It was a perfume, not toilet water, and she left no part of her body unanointed. For days after she left, the scent of that heavy French perfume filled her room.

Next she wanted towels. "I'm going to be unconscious," she said, "I hope, I hope, and the kidneys are going to work overtime trying to get rid of the poison. There's no point ruining a perfectly good, almost new mattress."

With my help she put four thick, folded bath towels beneath her hips. My dear Jesus Christ! Farewell to our bodies, but let no harm come to our possessions.

Next came Carmen's daily substitute for life's ordinary thrills: were the lethal pills where we had hid them? Yes, certainly. Goldie hadn't been permitted to know of their existence, and who else was there to hide them from? Since they were put in a new and supposedly more secure place every other day, the panic was occasioned not by the chance of their being stolen, but by the chance of our being unable to remember where we had put them.

I remembered this time. I had put them in the silver-mesh evening bag Carmen had given me; and the bag was with the other things of Carmen's I was to take home. The pills were in a big, square, heavy glass bottle, shaped a good deal like L'Interdit. When she saw them, Carmen smiled. "I was afraid that what we really intended to do was to lose them completely."

I heard Bill, who had been out to a Chamber of Commerce dinner, come into the kitchen. He would pause there, I knew, for a drink.

"Don't get him," Carmen said. "He doesn't want to be here when I take them and I don't want him to be." Bill had had pneumonia and two heart attacks during the past year. "He can't take it. I could probably get them down myself, but you stay. I don't want Bill here, keeling over when the job's half-done."

In the beginning, Carmen had said that she must be alone when she swallowed the pills. She had changed her mind. Not out of sentiment, no "We two together at the last," but "I might chicken-out if you weren't here, and I hadn't told you a hundred times what I was going to do."

I opened the sliding glass doors that led from Carmen's room to her patio with its low adobe walls: *I* was saying a good-by once again to the world from which Carmen had already parted. The sky was overcast, and a bird, a mockingbird, I suppose, was singing in what the cleaning lady called the "prickle bush." I shut the doors. When they were opened again, Carmen would be carried out and the bird would still sing in the prickle bush. A definition of life: prickle bushes in which birds sing.

I counted the pills out into a piece of Kleenex, Carmen watching and keeping count.

"What's the extra one for?" Carmen asked.

"Luck," I said.

But when I took the pills to her in their hobo bag of Kleenex, Carmen got off her carefully prepared pallet of towels and went quickly, silk panties and silk nightgown swishing, to her dressing table. There she emptied the cut-glass, gold-rimmed little jar intended, when Grace had owned it, for toothpicks, but filled now with matches, and brought it to me.

"Put the pills in this," she said.

My God, yes. Who wants to take his death from a crumple of Kleenex? What was the hemlock served in? Carmen was going to die as she had lived, with some attention to appearance.

Carmen's first decision to be alone when the pills were taken may have been the right one, for next I said what I would give a piece of my tongue at the very least to forget having said.

"They put you in my arms when you were born," I began, and Carmen's eyes filled with tears.

"Drink this, drink this," I said, holding the big glass tumbler of water. "You cannot swallow and cry."

She drank and was immediately calm.

She swallowed one capsule.

"Can you take two at a time?"

She answered with bravado. "I sure can."

She sure could. The small jar, ruby-colored with capsules, was soon its original color.

She lay back against the pillows. "I don't believe this will do it."

"The book I read, the doctors up north, said they would."

"I don't feel a bit different."

"It's not strychnine. Give it a little time."

"We'd about as well watch something while we wait."

I turned on "International Show Time," a kind of TV vaudeville program we both liked.

This would not have been Grace's way; or the way of any of Grace's forebears. Death by vaudeville. But Carmen wasn't casual about death. In the notebooks she had given me, she had written when eighteen, "Clifford Allee believes that when we die, we die, and are dead for good. I can't agree with him. I think there is an afterlife . . . there must be. I wouldn't like to die, but I don't believe I'd be especially frightened. I might even be the first one to run. Who knows?"

Three months before, when only seventeen, she had written, "I hate pain. I wouldn't give two cents if I kicked the bucket tonight. I don't care. I don't care. I don't care."

Which best expressed her convictions? Both? "I hate pain." And, "I think there is an afterlife."

"Don't you want me to get Bill?"

"Yes," she said.

Bill was at the kitchen bar, glass in front of him.

"The pills are down," I told him. "She's all right. Watching TV. She wants you."

Bill came, sat in a chair beside the bed. I, on the other side, sat on the twin bed.

"Turn off the TV," Carmen said.

While I was doing this, Carmen and Bill spoke together and I was glad that I couldn't hear their words. When I came back, Carmen said, "I still hurt. There's no point in that, is there?"

"None in the world."

"Or the next," Carmen said.

I gave her a shot, just the usual, no stronger. More seemed unnecessary, a breach of faith with all I had learned, overkill in the exact sense of those words. The shot took immediate effect. I was holding one of Carmen's hands, Bill the other. She lay back against her pillows, the frown of pain gone from her face. Then, smiling, her tone fond and mocking, her eyes moving from one of us to the other, she spoke a long sentence, perhaps two or three. She thought that what she said was amusing. She half laughed herself; no, not laughed, but smiled, expecting laughter from us; having said something gleeful and ironic.

We could not understand a word she spoke. Her brain still functioned, she chose the words with which to part from us, but her tongue was no longer able to obey the brain's commands.

I do not know what the expression of our faces conveyed to her. Not, I hope, our bewilderment. Drugged, dying, she was trying very hard to make the farewell gay and light. No crying because the party was over. I think she hadn't the slightest idea that the words she spoke were unintelligible. If we looked puzzled, I hope she attributed it to our slowness.

Her eyes closed at once. Bill left the room. It had been our plan for me to take a few sleeping pills as she took her many. We had both forgotten about it, and I didn't feel like sleeping.

I lay down on the bed next to hers. Carmen had stretched out a hand to help me straighten the covers of my bed after her decision had been made. I lay on the bed she had tidied for me as she lay on the bed she had chosen for herself.

Sleep was impossible. I was wide-awake, calm, not brokenhearted. Carmen appeared to be sleeping much as usual; perhaps her breathing was heavier. Death is absence, but absence itself cannot be accepted as an unchanging condition for some months. The footstep, the voice, the scent to which years have accustomed you assail nose and ears for weeks after death. Only time, which accustomed you to the quality of one person's life, can convince you that that life has ended.

I got off the bed and wandered about the neat, almost empty room. I had read that patients anesthetized for operations responded, though unconscious, to words of encouragement from their doctors and nurses. I put my arms around Carmen, my face next to hers and said, "You are doing fine. You are sleeping. You won't wake up. The pain is gone forever."

Carmen had taken pleasure in the irony of recording the time for her shots in an appointment book called "When and Where." What *she* wrote in it was "Earliest time for next shot." That gave her courage.

"Only so many more minutes to endure." I recorded the actual time of the last shot.

Carmen, fearing I know not what investigation, had asked me to keep the record, as we had been doing, until her death. She wanted no discrepancies to imperil Bill or Munger. It was a foolish fear, I thought, but, sleepless, I was glad to have some occupation.

In the drawer of her night table, behind her "When and Where" book, her emery boards, pen, glasses, was a semi-stiff square of paper on which Carmen had copied a quotation from a letter I had sent her after her report of "a recurrence, inoperable."

The quotation was from one of the minor English authors I had discovered on my afternoon tours of London's secondhand bookshops: Hesketh Pearson. He had written and I had copied out his words for Carmen; and she in turn had taken the quotation from my letter and put it in a more convenient form for her reading.

Hesketh Pearson had written, "Incidentally, it may comfort those who fear death to know from one whose life has three times been despaired of that while hovering on the verge of eclipse I did not review my past life like Queen Victoria in Lytton Strachey's book; nor was I in the least concerned over 'my future.' Such matters are only of moment when one has strength and leisure to reflect upon them. I simply wanted to be allowed to die as quickly as possible, to go to sleep and never wake up; and each time on being recalled to life my first feeling was one of resentment. Yet no one has been happier than

I have been and no one has loved life more than I."

I had, without knowing it, perhaps been remembering Pearson's words when Carmen and I talked of death when she was with me in July. Then I said, "You can say, 'It is not now,' and be happy. And you can say, 'When it comes, I'll be glad.'"

I put the card with Pearson's words in the silver-mesh bag along with the now half-filled bottle of pills with its promise of the sleep Pearson had wished for and Carmen had found.

Still unable to sleep myself, I walked about the big, tidy room. In the closet hung one dress, the one Carmen wanted to be buried in. I had never seen her wear it. Was it a new one bought for this purpose? Was it a dress she had had, had been happy in, and wanted for that reason to wear once more? Did she choose it because it would blend with the earth? It was green as grass, sprigged with flowers, simple and flowing as a waterfall.

I will never know the exact reason she chose it, other than it was becoming to Carmen and suitable, I thought, for its purpose. Why go to earth in something that contradicts earth and its products? The dead body is no flower or fruit; but like flowers and fruits, it returns to earth and has no more need to wear a pin-striped suit or a double-knit, self-belted chemise than has a persimmon or hummingbird. Carmen perhaps had not reasoned this way; but she had chosen with her sense of fitness a dress like one Ceres or Persephone might have worn—had they been mortal.

She did not care where she was buried. Gone was gone. Dead was dead. She had been willing to occupy the grave site intended for a brother's son; a site useless now that the son would be buried with his wife and children.

After a while I stretched out, still clothed, on the bed Carmen, with her already drugged hands, had helped spread for me. I slept as soundly as she—perhaps more soundly. She perhaps dreamed. I could remember no dreams. I awakened at the hour when I usually gave her her last shot before Goldie arrived.

Drugged and sleeping, that earlier rhythm still perhaps ruled Carmen, too. She was less quiet than she had been. Her face was drawn as it had been on those mornings when she felt pain. I remembered what she had said the night before. "I hurt." Why hurt? What was to be gained by hurting now?

I gave her the usual morning shot of Numorphan. She relaxed at once, appeared to be asleep and no longer hurting.

It was a bright, sunny day, and very long. Goldie had the night before been given the day off. Bill had meetings downtown and thought it best that on this day, he be seen at those meetings.

One does not, Pearson said, relive when dying one's past life. But I was not dying, and I relived as much of Carmen's life as I knew.

We were once in an earthquake together. I ran downstairs and out of the house, pell-mell. From an upper window, Carmen called to me as the house rocked, "Hey, you don't know what you're missing. It's like a cradle."

When in pain, or after having a painkiller shot, her hazel eyes darkened to a hickory color.

She was one or two and I five or six when I had to push her in a perambulator to make her drowsy. After just so much pushing and no drowsiness, I would pinch her so that I could tell Grace, "She fusses, no matter what I do." But she, as if knowing my purpose, would not cry; and I, soft-hearted, was incapable of giving her a pinch that would produce a cry in spite of her resolution.

She could pretend a pain she did not feel if the occasion demanded it. "My stomach hurts," she would tell Grace; and Grace became convinced that Carmen had inherited her grandmother's weak stomach and would say, "Jessamyn and I will do the dishes." It was a piece of acting of which Carmen was proud. "I never felt a thing," she told me later.

Never once during the time I was with her during her illness did she bewail her fate, berate herself for not seeing a doctor sooner, or criticize her doctor for not insisting on more radical surgery.

She endured. She planned. She carried out her plan.

I made frequent trips to her bedside to see Carmen; but I spent most of the forenoon of that day in Carmen's cactus garden, as she called it. It was now October 26, Christmas less than two months away; but the desert sun was still hot.

About noon Carmen's breathing changed. Before then it had been two short heavily indrawn breaths and one long expelled breath.

I was with her when the change came. Her face was still pink, lipstick unsmudged, hair undisturbed. She looked less drugged than many a time after a heavy shot. She was warm, did not feel to be feverish, was still Carmen. I thought that if I lifted her in my arms, said, "Wake up, Carmen, it's getting late," she would open her eyes, say, "What time is it?"

Up to that moment she had seemed to be sleeping. Then her breathing changed. She began to breathe like a baby who has the croup, great dragging watery breaths. Now her lungs were filling with water, now she was drowning. So long as she appeared to be sleeping, there was no inclination to awaken her; but drowning was a different matter. Shouldn't I try by some means to save her?

Save her for what? Save her for months of pain until the lungs once again didn't function properly?

I knelt by her bedside. I said, "This is the hard part, Carmen. Your body is so strong it fights to save you. It fights like a machine. It doesn't know what it is saving you for. The lungs will give up pretty soon. Don't worry. Take it easy. Tell your lungs you've had all the air you want."

I spent most of that day outside. Once a car drove into the driveway, a friend bringing what she called "an idiot soufflé" to heat up for Carmen's dinner. She could hear Carmen's breathing.

"She's had a shot. She's sleeping. The medication affects the lungs," I told her.

The friend left at once. It wasn't a pleasant sound to hear.

I put the soufflé on the sink, went through Carmen's room on my way to my sun-tan cot.

Outside, I thought of the past weeks, and of the persistent tone of humor in Carmen's voice.

One night, when I awakened her to give her a shot, she spoke in a voice light and wondering, very far from complaint.

"I was dreaming about our Republican ancestors."

"What were you dreaming?"

"I was dreaming that they were having a hearti-choke."

"A *what?*"

"You know. A zannibar."

Now I think she knew she was saying the wrong word, her voice was so mischievous.

"What's that?"

"You know. Where they all keep talking so that a bill can't be passed."

Carmen had a phone on her night stand from which she could call Bill in his room. "Call Bill," I said. "If it's a political matter, he'll know the right name for it."

The idea of rousing Bill from his sacrosanct sleep made her laugh.

"Filibuster," she said, as if she had known the right word all along.

The only other dream she ever told me about was of the kind she said sleeping pills on top of Numorphan gave her: a nightmare. She dreamed that she was in an automobile with Grace, the car driven by someone else, and that Grace was trying to push her out.

"She was trying to kill me."

"Why would she do that?"

"She was trying to punish me for taking more than my fair share of the rabbit-sack dishcloths she had made when we divided up her things."

"Mama was dead then. How could she know how many you took?"

"That's what made the dream so bad. Mama was dead and still she knew and wanted to kill me for being so selfish."

"She didn't know and she didn't want to kill you. That dream was the result of a combination of Numorphan and Percodan and Parnate and Seconal and God knows what else."

"Do you think so?"

"Of course."

"And my bad conscience?"

"Maybe. And I can ease that by taking those extra dish towels home with me when I go. No use leaving them here for one of those attractive ladies you're having your friends line up for Bill. Is there? You don't want him courted for his dish towels, do you?"

Carmen laughed.

How *could* she laugh? Let me be happy playing the clown when she was dying? And I remembered telling Grace, a sodden, lugubrious weight on her already burdened shoulders, "Even if I get well, I'll never smile again." Thank God I didn't have a sick me to look after.

The sunny day of October 26 began in late afternoon to cloud over. In the west, the sun peered through bars of gray-black clouds like a tiger in a cage. It was time to go in, before the desert cold filled the valley.

Carmen was lying as she had been all afternoon. Eyes closed, her breathing deep, heavy, and soggy. I went into her bathroom, looked in the mirror as if I expected to see reflected in my face something of what Carmen was going through. While I was looking, searching for the shadow I thought should be there, my eyes no longer functioned. I could not see. I could only hear—and what I heard was silence. After twenty-four hours, the human machine that had struggled so hard to survive intolerable punishment had ceased to function. The lungs no longer struggled to bring in air.

Carmen was dead.

She was still warm with life. Her hand was still a live hand in mine, flexible, almost, it seemed, clinging.

"My darling, you did it. You had the courage. It was a great gift to all of us. To depart like a courteous guest. You did not wear out your welcome. You did not linger to cause us all to suffer. All as you planned. Nothing more to worry about now."

I did not call Dr. Munger. I left that for Bill, and I couldn't call Bill. I didn't know where he was.

Dr. Munger came immediately after Bill's call. He did not once look at Carmen, whose face was gray now; and I knew that she would be glad for that. Munger kept his eyes on his stethoscope, which he placed over her chest.

When he had finished listening, he turned to us. "What happened?"

Bill, by now coached, answered, "She was sitting up, talking, in good spirits, then she suddenly slumped back onto her pillows."

"Sounds like a pulmonary embolism," Dr. Munger said. "Since the cancer would have killed her anyway and much more painfully, we can only be thankful. She was a wonderful woman."

To Bill he said, "Would you like me to call the undertaker?"

"Please," said Bill.

To me he said, "Get those towels out from under her. She made a big effort to be neat."

I did. I put them in the incinerator; and stayed

outside until the undertaker had come and had taken Carmen's body out of her room and placed her in his long black car. This was something I didn't want to see.

I went down to the Funeral Home, as it was called, late the next afternoon. Mr. Newsom, who owned it, met me.

"May I see Carmen?"

"Your sister phoned me about you a month or two ago. The casket was to be opened for no one else, she said. But she was sure you would be down and she wanted it opened for you."

The casket, as pretty a gray box as Bill could have found for this purpose, was in a room named "Repose" or "Rest" or "Prayer." Something like that.

Death, erasing as it had the lines of pain from Carmen's face, made her look ten years younger. Her hair had been properly done by someone who knew how Carmen wore it. The green dress was like a leaf, fallen early.

"Little sister, little sister. I can't make you laugh any more. But I don't have to worry about making you cry any more, either. They did put you in my arms when you were born—and here we are together at the last. I love you, little sister."

Then I read the words James Baldwin had written for the death of Rufus, words she had hoped would be said at her funeral.

"You know a lot of people say that a man who takes his own life oughtn't to be buried in holy ground. I don't know nothing about that. All *I* know, God made every bit of ground I ever walked on and everything God made is *holy*. And don't none of us knows, what goes on in the heart of some one, don't many of us know what goes on in our own hearts for the matter of that, so can't none of us say why he did what he did. Ain't none of us been there so we don't know. We got to pray that the Lord will receive him like we pray that the Lord's going to receive us. That's all. That's *all.*"

I said those words as Carmen lay in her green dress in the gray box. The words she had wanted said at her funeral.

They weren't said there. They were the last words I ever spoke aloud to Carmen; though in my mind, silently, I speak to her almost every day.